Lab Manual
to Accompany

Introduction to

BIOTECHNOLOGY

AN AGRICULTURAL REVOLUTION

Lab Manual
to Accompany

Introduction to

BIOTECHNOLOGY

AN AGRICULTURAL REVOLUTION

RAY V. HERREN

DELMAR
CENGAGE Learning®

Australia • Brazil • Japan • Korea • Mexico • Singapore • Spain • United Kingdom • United States

DELMAR
CENGAGE Learning

Lab Manual to Accompany Introduction to Biotechnology, 2E
Ray V. Herren

Vice President, Editorial: Dave Garza
Director of Learning Solutions: Matt Kane
Senior Acquisitions Editor: Sherry Dickinson
Managing Editor: Marah Bellegarde
Senior Product Manager: Christina Gifford
Editorial Assistant: Scott Royael
Vice President, Marketing: Jennifer Baker
Marketing Director: Debbie Yarnell
Marketing Coordinator: Erin Deangelo
Senior Production Director: Wendy A. Troeger
Production Manager: Mark Bernard
Senior Content Project Manager: Katie McGuire
Senior Art Director: Dave Arsenault

Library of Congress Control Number: 2011941347

ISBN-13: 978-1-133-13069-7

ISBN-10: 1-133-13069-0

Delmar
Executive Woods
5 Maxwell Drive
Clifton Park, NY 12065
USA

Cengage Learning is a leading provider of customized learning solutions with office locations around the globe, including Singapore, the United Kingdom, Australia, Mexico, Brazil, and Japan. Locate your local office at **www.cengage.com/global**

Cengage Learning products are represented in Canada by Nelson Education, Ltd.

To learn more about Delmar, visit **www.cengage.com/delmar**

Purchase any of our products at your local bookstore or at our preferred online store **www.cengagebrain.com**

Printed in the United States of America
2 3 4 5 6   22 21 20 19 18

Table of Contents

Name______________________________ Date______________

CHAPTER 1
The Phenomena of Biotechnology

LAB ACTIVITY 1.1
Global Use of Transgenic Crops

Purpose

The purpose of this lab is to have you identify the role that transgenic crops are playing in the world's agriculture. You will examine the amount of acreage devoted to, and the type and amount of crops grown for, genetically altered plants between 1996 and 2002.

Materials

- Graph paper or spreadsheet computer application
- Colored pencils
- Ruler

Procedure

1. Use the data on the acreage of transgenic crops shown in Lab Table 1–1 to construct a bar graph. The *x*-axis should be labeled "year" and the *y*-axis should be labeled "millions of acres." Use a different-colored pencil for each bar and label each bar with the specific acreage for that year.

LAB TABLE 1–1 World Acreage of Transgenic Crops

Year	Acres (millions)
1996	4.3
1997	27.5
1998	69.5
1999	98.6
2000	109.2
2001	130.0
2002	146.3

Name ____________________ Date ____________

2. Next, construct a multiple line graph that shows the specific types of transgenic crops planted between 1996 and 2002, as shown in Lab Table 1–2. The *x*-axis should be labeled "year" and the *y*-axis "millions of acres." Use a different-colored pencil for each line and create a key for your graph.

LAB TABLE 1–2 Types of Transgenic Crops Grown, 1996–2002 (millions of acres)

	1996	1997	1998	1999	2000	2001	2002
Corn	0.8	8.0	20.8	27.8	25.8	24.5	31.0
Soybeans	1.3	12.8	36.3	54.0	64.5	83.3	91.3
Cotton	2.0	3.5	6.3	9.3	13.3	17.0	17.0
Canola	0.3	3.0	6.0	8.5	7.0	6.8	7.5

3. The last graph you will create will be a stacked bar graph that shows the amount of global acreage devoted to transgenic crop production compared to total acreage planted for four different crops, as shown in Lab Table 1–3. The *x*-axis should be labeled "year" and the *y*-axis "millions of acres." Use a different-colored pencil for each bar and label each bar with the specific acreage for that year.

LAB TABLE 1–3 Global Acreage of Transgenic Crops Compared to Traditional Crops Planted in 2001 (millions of acres)

	Transgenic Crops	Total Crops
Corn	24.5	350.0
Soybeans	83.3	180.0
Cotton	17.0	85.0
Canola	6.8	62.5

4. Use all three of your graphs to complete the conclusion exercises.

Name ______________________________ Date ______________

Conclusions

1. Describe the trend in the number of transgenic crops planted between 1996 and 2002.

2. Using the bar graph constructed from Lab Table 1–1, calculate the annual rate of increase in millions of acres for transgenic crops planted around the world between 1996 and 2002. Rate of increase can be calculated by using the formula: rate of increase = change in value/change in time.

3. Using your graphs, identify which two transgenic crops make up most of the genetically altered crops grown around the world.

4. Using the stacked bar graph constructed from Lab Table 1–3, calculate the percentage that each transgenic crop occupies out of the total amount of crops grown around the world.

Name ____ Date ______________

LAB ACTIVITY 1.2
Making Buttermilk

Purpose

The purpose of this lab is to introduce you to one of the earliest uses of biotechnology—making fermented milk products using bacteria.

Materials

- Fresh cultured buttermilk
- Quart jar with lid
- Whole milk
- Permanent marker or oil pencil
- Refrigerator

Procedure

1. Make sure your hands and glassware are clean and dry. Sterilizing glassware before starting the lab is preferable.
2. Add 8 ounces of fresh cultured buttermilk to a clean, dry quart jar.
3. Fill the jar to the top measure line with fresh whole milk.
4. Screw the lid on the jar tightly and invert the jar to ensure that the milk and buttermilk mix thoroughly.
5. Label the jar with your name and the date and time.
6. Make observations of the appearance and thickness (viscosity) of the liquid in the jar.
7. Place the jar in a warm location (room temperature or slightly higher) and let sit overnight.
8. Make observations of the appearance and thickness (viscosity) of the liquid in the jar after 24 hours.
9. Place finished buttermilk in the refrigerator for storage.

Name______________________________ Date______________

Conclusions

1. Compare the initial appearance of the liquid in the jar with the appearance of the liquid after 24 to 36 hours.

2. What was the purpose of adding the 8 ounces of buttermilk to the jar?

3. What process converts whole milk into buttermilk?

4. Why is the transformation of milk into buttermilk considered a form of biotechnology?

Name____ Date______________

CHAPTER 2
The Historical Development of Biotechnology

LAB ACTIVITY 2.1
Biotechnology Time Line

Purpose

The purpose of this lab is to have you identify the important milestones in the development of biotechnology.

Materials

- PowerPoint presentation software or poster paper
- Colored pencils or felt-tipped pens

Procedure

1. You are going to create a time line that illustrates the advances and highlights in the field of biotechnology over its long history. Your instructor will decide if you are going to do your time line on the computer using PowerPoint or as a large poster.
2. You first have to research the history of biotechnology. This can be done by using the library or the Internet. Use the topics listed in Lab Table 2–1 to help you get started in your research. They are not listed in chronological order—it is up to you to identify the time frame for these topics. They are meant only to help you begin your time line, so do not limit your research to just these topics. Your instructor will help you in expanding the topics you will use in building your time line.
3. Once your research is complete, you must then create your time line in the format your instructor has chosen. Try to be creative—use pictures whenever possible. Also make sure to highlight the specific years when certain advances were made in the field of biotechnology. Arrange your time line chronologically, starting with the oldest developments first and leading up to the present day.
4. When your time line is finished, use it to complete the conclusion exercises.

Name ______________________________ Date ______________

LAB TABLE 2–1 Research Term for the Biotechnology Time Line

Fermentation	Green Revolution
DNA	Genes
Plant breeding	Insulin
Yeast	Transgenic plants
Genetics	Genetic engineering
Antibiotics	Bt cotton
Vaccine	Flavr-Savr tomato
Artificial insemination	Vinegar
Beer	Yogurt
Bread	Enzyme
Wine	William Beal
Cheese	Bioremediation
Gregor Mendel	Cloning
In vitro culture	Golden rice
James Watson	Roundup Ready
Francis Crick	Wastewater treatment
Louis Pasteur	

Name ______________________________ Date ______________

Conclusions

1. What was the first use of biotechnology you discovered in your research, and how long ago did it occur?

2. What was the most recent advance in biotechnology you discovered in your research?

3. Describe some of the problems that biotechnology has solved through time.

4. Which advance in biotechnology do you think had the greatest impact on human society, and why?

Name____ Date ________________

LAB ACTIVITY 2.2
Making Yogurt

Purpose

The purpose of this lab is to introduce you to one of the earliest uses of biotechnology—making fermented milk products using bacteria. This lab may be used in conjunction with the making cheese lab from Chapter 3.

Materials

- 1 gallon whole milk
- Plain yogurt (as fresh as possible)
- Cooking pot with heavy bottom or double boiler
- Heat source (stove or hot plate)
- Thermometer
- Sterilized glass jars (quart size)
- Large spoon or whisk for mixing
- Measuring cup (2-cup capacity)

Procedure

1. Before starting the lab, be sure that your hands are clean and dry and all glassware has been sterilized.
2. Place 1 gallon of whole milk into the cooking pot.
3. Place the cooking pot on the heat source and heat slowly over medium heat. Stir milk so that it does not burn to the bottom.
4. Heat the milk until it is 185°F to 195°F (85°C to 90°C). Do not let the milk boil over.
5. Place the pot of milk in a tub of cold water and cool to 120°F to 130°F (50°C to 55°C).
6. Place 1 cup of the milk into the measuring cup and add 1 cup of fresh plain yogurt.
7. Stir to blend the yogurt and milk until the mixture is uniform.
8. Pour the milk and yogurt blend back into the pot of milk slowly stirring continuously. Mix well.
9. Pour the milk/yogurt mix from the pot into the sterilized jars.

10. Place in an incubator set to 125°F (50°C). (You can also use a cooler filled with warmed water in place of an incubator.)
11. Let the jars with the mixture sit for 3 hours. At this point the yogurt should have gelled.
12. The yogurt may be mixed with sugar and fresh or frozen fruits and tasted by students. If there is any doubt about the quality of the yogurt (odd appearance or smell), *do not eat it!*
13. Place the jars in a refrigerator for storage.

Name ____________________ Date __________

Conclusions

1. Why did you add fresh yogurt to the heated milk?

2. Why did you use heated milk rather than cold milk?

3. Fresh milk can last for only a few days without spoiling, and yogurt may last for 3 weeks. Why do you think this is possible?

4. How does the process used to create yogurt compare to the process used to create buttermilk in the lab from Chapter 1?

Name____________________ Date__________

CHAPTER 3
The Principles of Scientific Research

LAB ACTIVITY 3.1
Zymotechnology: Ancient Art and Modern Science

Purpose

The purpose of this lab is to introduce you to the process of fermentation by yeast, a single-celled fungus that consumes sugar and produces carbon dioxide gas and alcohol. This process is probably the oldest use of biotechnology. The fermentation process is the same process that bakers have used for thousands of years to make bread rise. The use of yeast as a technology is also known as zymotechnology.

Materials

- One 12-ounce soda or juice bottle
- Large balloon
- 1 teaspoon dry yeast?
- 1 teaspoon sugar
- Measuring spoons
- Measuring cups
- Masking tape for labeling
- Stopwatch
- 1 cup warm water

Procedure

1. Collect all the required materials and bring them to your work area. Mix 1 cup of warm water with one teaspoon of sugar and 1 teaspoon of dry yeast in a clean soda or juice bottle.
2. Immediately after you have thoroughly mixed the ingredients in the bottle, cover the top of the bottle by stretching a balloon over the opening. Make sure you carefully pull the balloon down far enough over the top of the bottle to ensure an airtight seal.
3. Using the masking tape, label the bottle with your group name and set it aside in a warm, dark area. The yeast should begin converting the sugar into carbon dioxide almost immediately. Begin timing how long it takes to fill the balloon with carbon dioxide gas. The carbon dioxide will begin to fill up the balloon, much like the way the gas collects in bread dough and causes it to rise. When the balloon is completely filled, carefully let out the gas by lifting the side of the balloon off the top of the bottle. When the balloon is deflated, cover the top again and observe if more carbon dioxide gas is generated.
4. When the experiment is finished, complete the following conclusion exercises.

Name________________________________ Date______________

Conclusions

1. How long did it take for the balloon to completely fill with carbon dioxide gas?

2. Write out the chemical reaction for fermentation.

3. Describe how the experiment you just conducted illustrates why the process of fermentation is considered biotechnology.

Name____ Date________________

LAB ACTIVITY 3.2
Making Cheese

Purpose

The purpose of this lab is to introduce you to one of the earliest uses of biotechnology—making fermented milk products using bacteria.

Part I—Making Soft Labneh (Lebanese Origin)

Materials

- 1 quart of yogurt
- 1 teaspoon salt
- Cheesecloth (sterilized)
- Two bowls
- Whisk
- Strainer or large funnel
- String
- Refrigerator

Procedure

1. Place 1 quart of yogurt into a bowl and stir until smooth.
2. Add 1 teaspoon of salt to the bowl. Use the whisk to mix well.
3. Line the strainer with the cheesecloth and place it over a clean bowl.
4. Pour the salted yogurt through the cheesecloth.
5. Lift the corners of the cheesecloth to contain the solid part of the yogurt (curds) and let the remaining liquid (whey) drip out.
6. Suspend the cheesecloth containing the curds over a bowl in the refrigerator. Let sit for 24 hours to allow all the whey to drain. You can try to speed up the process by *gently* squeezing the cheesecloth to remove additional moisture.
7. After the whey has finished draining, open the cloth and form the remaining curds into a ball. Wrap and store in the refrigerator.

Name ______________________________ Date ______________

Part II—Making Neufchatel (French Origin)

Materials

- 1 gallon whole milk
- ¼ cup cultured buttermilk
- ¼ tablet of rennin
- ¼ cup of water
- Large pot (5-quart capacity)
- Hot plate or stove
- Thermometer
- Ladle
- Large bowl
- Cheesecloth (sterilized)
- Strainer or large funnel
- Salt (optional)
- Refrigerator

Procedure

1. Sterilize the pot by boiling water in it for 5 minutes.
2. Place the milk and buttermilk in a pot.
3. Heat the milk and buttermilk mixture to a temperature of 65°F. Stir continuously. Use the thermometer to monitor the milk's temperature. Do not overheat the milk.
4. Dissolve the rennin in ¼ cup of water.
5. Add the water with rennin into the pot of milk/buttermilk. Stir to mix.
6. Cover the pot, and let sit overnight—or until a curd has formed—at room temperature (65°F to 70°F).
7. Use the ladle to remove the curds from the whey.
8. Place the cheesecloth into the strainer and place the strainer over the bowl. Place the curds in the clean cheesecloth.
9. Pour the remaining whey through the cheesecloth to remove any remaining curds.
10. Suspend the curds in the cheesecloth over a bowl in the refrigerator for 24 hours to let the remaining whey drain out. You can speed up this process by *gently* squeezing the curds in the cloth to remove liquid.
11. Turn the curds out of the cheesecloth and salt to taste (optional).

Name ____________________ Date ____________

Conclusions

1. Compare the consistency of the cheese from Part I of the lab with the consistency of the cheese from Part II of the lab. Do you notice any differences?

2. What was the purpose of the yogurt in Part I and the buttermilk in Part II?

3. What process creates the curds that form both cheeses?

4. What is the purpose of the rennin? Where is rennin found in nature?

5. Why did you heat the milk in Part II of this lab?

Name________________________________ Date______________

CHAPTER 4
Cells: The Foundation of Life

LAB ACTIVITY 4.1
Cells: The Building Blocks of Life

Purpose

In this activity, you will identify all the main parts of both plant and animal cells. Using a microscope, you will also observe live plant, animal, and bacterial cells.

Materials

- Colored pencils
- Microscope
- Tweezers
- Blank microscope slides and coverslips
- Toothpicks
- Eyedropper
- Water
- Lugol's iodine stain
- Methylene blue
- Onion
- Yogurt
- Elodea leaf

Name____ Date______________

Procedure A

1. Label all the parts of the animal cell shown in Lab Figure 4–1. Use a different-colored pencil to shade in each distinct cell part.

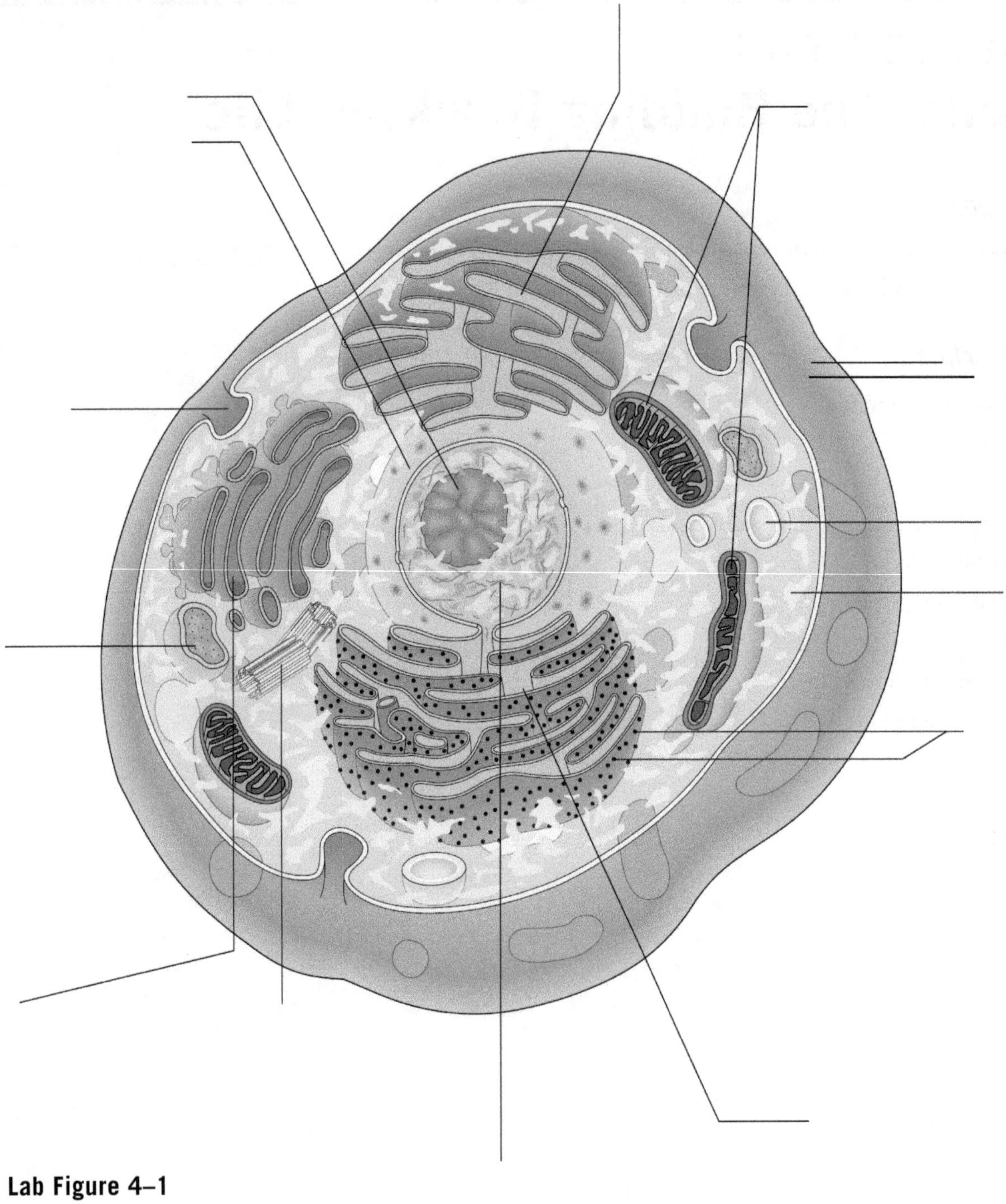

Lab Figure 4–1

Name____ Date______________

2. Label all the parts of the plant cell shown in Lab Figure 4–2. Use a different-colored pencil to shade in each distinct cell part.

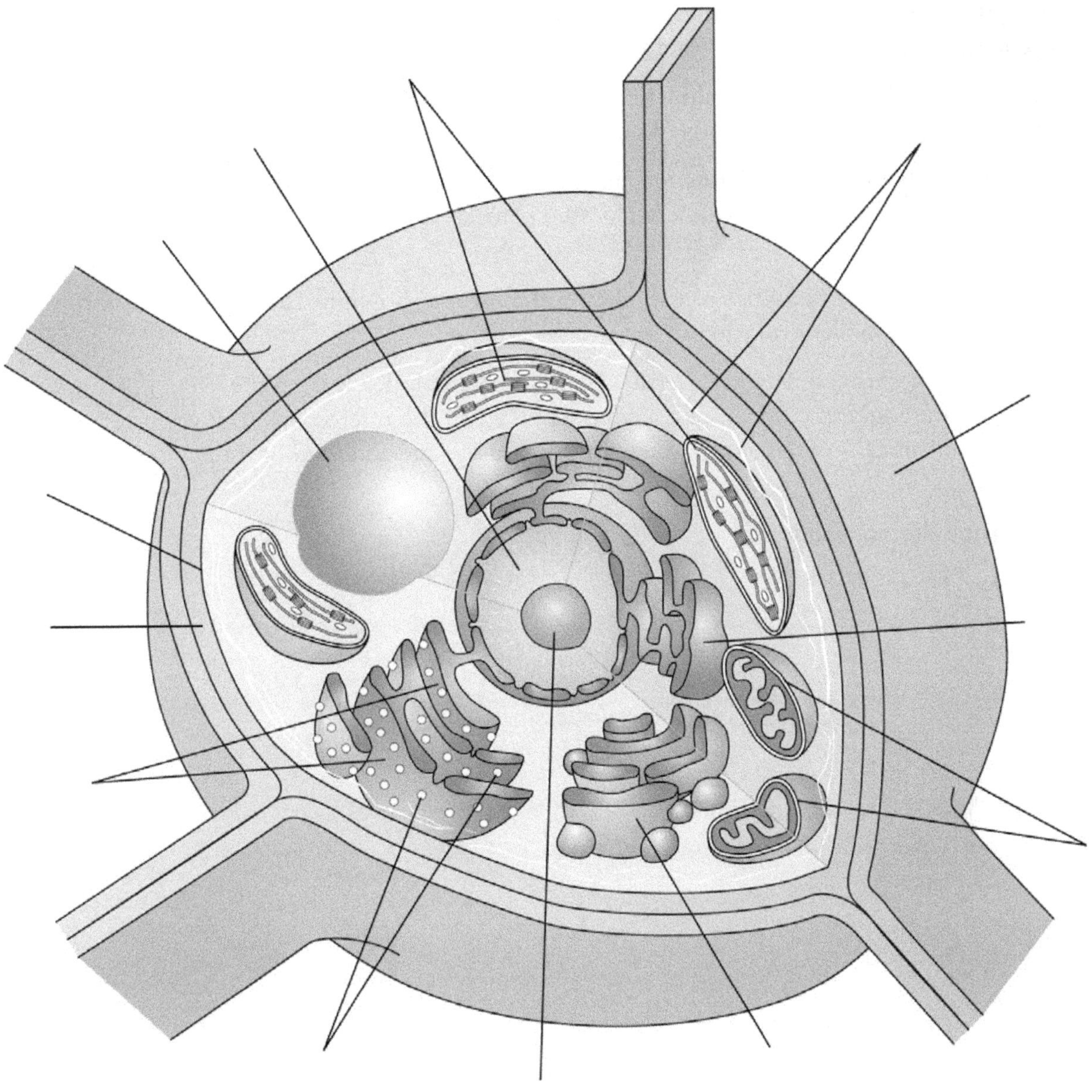

Lab Figure 4–2

Name______________________________ Date______________

Procedure B

1. First you will observe a live plant cell. Take an inner part of the onion and, using the tweezers, carefully peel off a thin section of the onion's inner skin. Place the section of skin on a blank microscope slide. Make sure the section lies flat and is not wrinkled or curled up. Use the eyedropper to place a drop of water on the skin and cover it with a coverslip.
2. Next, place a drop of iodine stain near the edge of the coverslip. This will stain the cells of the onion, making it easier to observe their organelles.
3. Place the slide under the microscope and observe the onion cells. Always begin at the microscope's lowest power. Once you have located some cells, carefully increase the power of the microscope.
4. Draw a rough sketch of what you see. Using the format shown in Lab Figure 4–3, also list all of the cell's organelles that you can see.
5. Clean and dry your slide and coverslip. Now place an elodea leaf on the slide. Put a drop of water on the leaf and cover it with a coverslip. Return the microscope to low power and observe the plant cells of the elodea. Sketch what you see and add to your list any organelles that are visible.
6. Clean and dry your slide.

Name____ Date ______________

Cell type: ___________

Rough sketch

Observed cell parts

Cell type: ___________

Rough sketch

Observed cell parts

Cell type: ___________

Rough sketch

Observed cell parts

Lab Figure 4–3

Name____ Date______________

Procedure C

1. Now you will observe a live animal cell. Carefully take a clean toothpick and rub it along the inside of your cheek. Then put a drop of methylene blue near the center of a clean microscope slide. Rub the toothpick in the methylene blue to mix in the cheek cells. Gently cover the mixture with a coverslip.
2. Place the slide under the microscope and observe the cheek cells. Always begin at the microscope's lowest power. Once you have located some cells, carefully increase the power of the microscope.
3. Draw a rough sketch of what you see. Using the format shown in Lab Figure 4–3, also list all of the cell's organelles that you can see.
4. Clean and dry your slide.

Procedure D

1. Finally, you will observe a prokaryotic bacterial cell. Use a toothpick to place a small amount of yogurt onto the middle of a blank microscope slide. Add a drop of water to the yogurt and use the toothpick to gently mix the yogurt and water together. Carefully cover the mixture with a coverslip.
2. Place the slide under the microscope and observe the rod-shaped bacterial cells that live in the yogurt. Always begin at the microscope's lowest power. Once you have located some cells, carefully increase the power of the microscope.
3. Draw a rough sketch of what you see. Using the format shown in Lab Figure 4–3, also list all of the cell's organelles that you can see.
4. Clean and dry your slide.

Name ______________________________ Date ______________

Conclusions

1. List all the parts of the plant cell that you observed in procedure B.

___

___

___

2. What cell parts did the plant cells have but the animal or bacterial cells did not have?

___

___

___

3. List all the parts of the animal cell that you observed in procedure C.

___

___

___

4. Which of the cells did you find easiest to observe?

___

___

___

5. Which of the cells was most difficult to observe?

___

___

___

6. Describe the main differences between an animal cell and a bacterial cell.

___

___

___

Name____ Date______________

LAB ACTIVITY 4.2
Diffusion through a Cell Membrane

Purpose

The purpose of this activity is to explore how the semipermeable cell membrane functions to allow materials in and out of cells.

Materials

- Cornstarch
- Water
- Beakers
- Tablespoon
- Graduated cylinder
- Plastic sandwich baggies (*not* zip type)
- Iodine

Procedure

1. Each student (or lab group) should obtain a plastic bag, a beaker, a tablespoon of cornstarch, and 100 ml of water.
2. Place the cornstarch and the water in the plastic bag. Tie off the top of the bag. Dissolve the cornstarch in the water by gently shaking the bag.
3. Fill the beaker about halfway with water. Add 10 drops of iodine to the water in the beaker.
4. Record the color of the water in the beaker and the color of the water in the bag.
5. Place the tied off bag into the beaker.
6. Let the bag sit in the beaker for 20 minutes. After 20 minutes, again record the color of the water in the beaker and the color of the water in the bag.

Name____________________ Date__________

Conclusions

1. What changes did you observe in the water in the beaker? What changes did you observe in the water in the bag?

2. What was the purpose of the iodine in this lab?

3. What did the plastic bag represent in this lab?

4. How do you explain the results you saw?

5. Why do you think the water in the beaker did not turn bluish/black?

Name____________________________ Date____________

CHAPTER 5
The Principles of Genetic Transfer

LAB ACTIVITY 5.1
Extracting DNA

Purpose

In this activity, you will identify all the structural parts of the DNA molecule. You will also extract actual DNA from a living organism and observe the DNA using a microscope.

Materials

- Colored pencils
- Microscope
- Blender
- Tweezers
- Blank microscope slides and coverslips
- Eyedropper
- Strainer
- Glass stirring rod or pipette
- Kitchen knife
- Cutting board
- Coffee filter
- Small onion
- 2 cups hot water
- 1 teaspoon salt
- 1/4 cup liquid dishwashing detergent
- 1/8 teaspoon meat tenderizer
- Ice-cold isopropyl alcohol (rubbing alcohol)
- 25-ml graduated cylinder
- 2 large glass beakers (500 ml)
- Test tube
- Methylene blue

Procedure A

1. Label all the parts of the DNA molecule shown in Lab Figure 5–1. Use a different-colored pencil to shade in each distinct component of the DNA. Next, label all the parts of Lab Figure 5–2.

Procedure B

1. Next, you are going to extract DNA from an onion so that the DNA can be observed with a microscope. First, peel and carefully chop one small onion into very small pieces.
2. Now put the onion pieces into the blender with approximately 2 cups of hot water and 1 teaspoon of salt. Process the onion mixture on high speed for about 3 to 4 seconds.

Name______________________________ Date______________

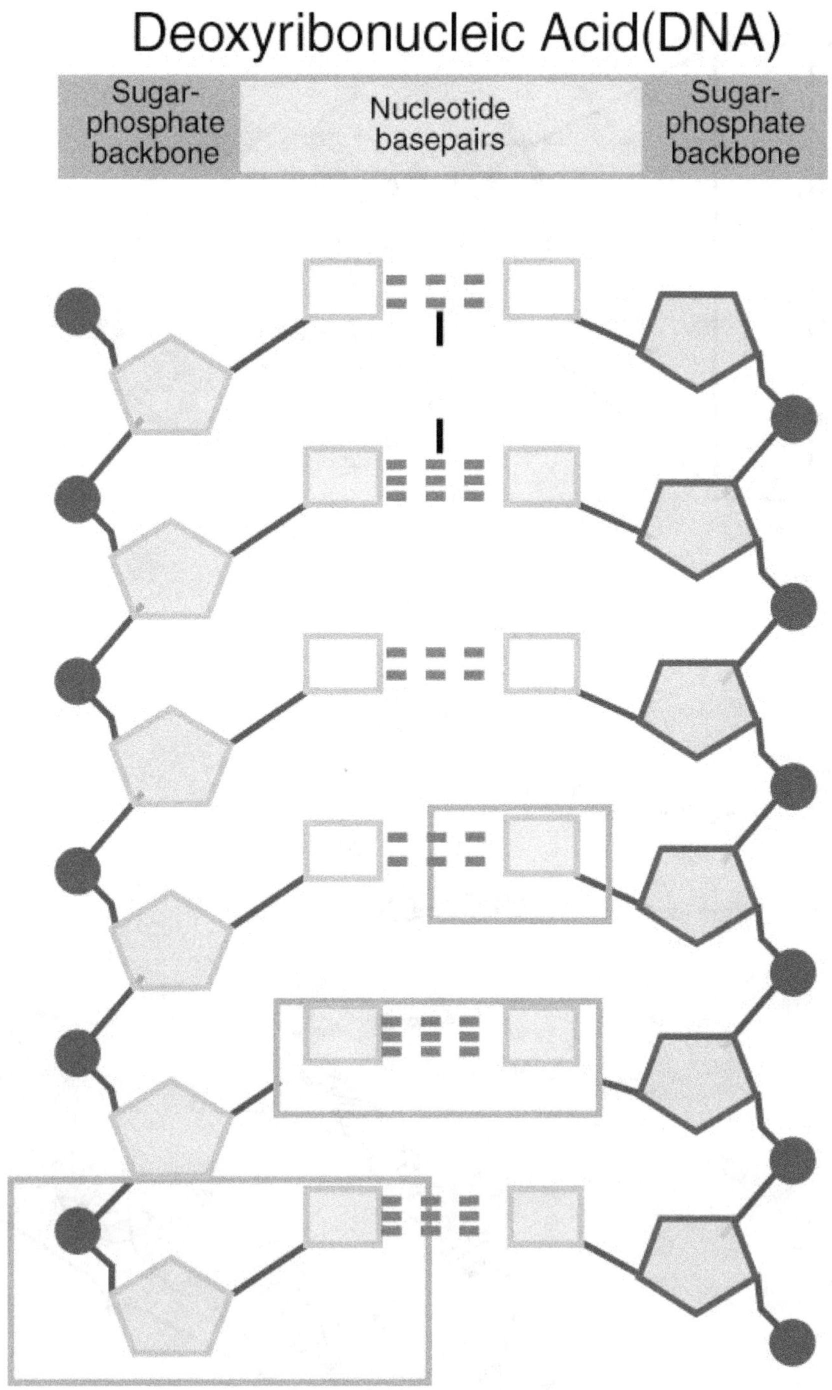

Lab Figure 5–1
Components of the DNA molecule

Name ____________________ Date ____________

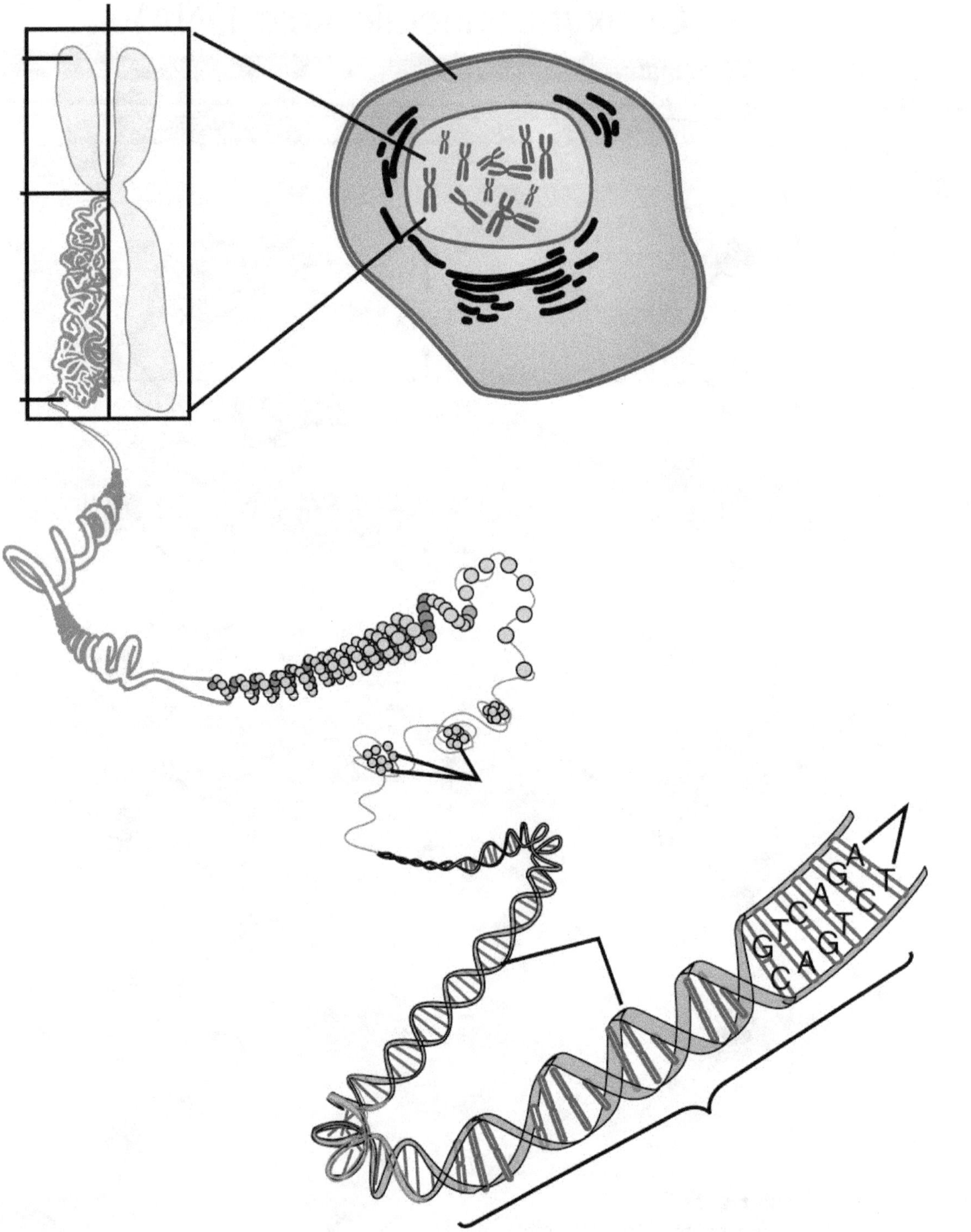

Lab Figure 5–2
Chromosomes

3. Pour the chopped onion mixture into a 500-ml beaker and add a little less than 1/4 cup of dishwashing detergent to the mixture. Stir the mixture with a clean glass stirring rod for 5 minutes.
4. Line the strainer with a coffee filter and hold it over an empty beaker. Then carefully pour the onion/detergent mixture into the strainer. This will help to separate the DNA from the mixture. Discard the filter.
5. Now add 1/8 teaspoon of meat tenderizer to the filtrate and stir with a clean glass rod for 5 minutes. This will help remove excess proteins from the filtrate.
6. Place an empty test tube in an empty beaker so that it sits at about a 30° angle. Carefully pour about 2 inches of filtrate into the test tube. Insert a clean glass stirring rod into the test tube.
7. Pour about 10 ml of ice-cold isopropyl alcohol into the graduated cylinder and slowly drizzle small amounts of the alcohol down the inside walls of the test tube. The alcohol should form a layer at the top of the filtrate, and stringy DNA molecules should begin to precipitate out of the solution and sink to the bottom of the test tube. Gently rotate the glass rod at the bottom of the test tube to spool up and collect the DNA.
8. Carefully remove the stirring rod from the test tube and transfer the DNA to a clean microscope slide. Add a drop of methylene blue stain to the DNA and cover it with a coverslip.
9. Observe the DNA by first viewing it with the lowest power of the microscope. Then gradually increase the power.
10. Clean up your lab materials.

Name ______________________________ Date ______________

Conclusions

1. Describe the function of DNA.

2. What are the four nucleotides that make up the DNA molecule, and which ones can bond with one another?

3. Explain the difference between DNA, genes, and chromosomes.

4. Explain why you had to put the onion into the blender.

5. What role did the dishwashing detergent play in this experiment?

6. Explain why DNA extraction is so important to the field of biotechnology.

7. Describe the appearance of the DNA you extracted in this experiment, with and without the microscope.

Name____ Date________________

LAB ACTIVITY 5.2
Observing Mitosis in Root Cells

Purpose

The purpose of this lab is to observe the different stages of mitosis in plant cells. Root tips are especially good for observing the stages of mitosis because root tips are constantly growing, so the cells are constantly dividing.

Materials

- Root tips (onion root tips work well)
- Razor blade or scalpel
- Petri dish
- Tweezers/forceps
- Microscope slides
- Coverslips
- Distilled water
- Droppers
- 1 molar HCl
- Crystal violet stain
- Paper towels
- Compound microscope

Procedure

1. Using the tweezers and the razor blade or scalpel, cut 1- to 2-mm pieces off the tips of the roots.
2. Place the root tips in the Petri dish and soak in several drops of 1 M HCl for 5 minutes.
3. Use the forceps to peel off a layer of the root tip.
4. Place a drop of distilled water on the microscope slide and place the root tip on the microscope slide.
5. Place the coverslip over the root tip and press down slightly until the coverslip lies flat.
6. Place one to two drops of crystal violet stain on the left-hand edge of the cover slip. Place a folded paper towel on the right-hand edge of the coverslip to draw the stain under the coverslip.
7. Let the stained slide stand for 2 to 3 minutes.
8. Place one to two drops of distilled water on the left-hand edge of the coverslip. Place a folded paper towel on the right-hand edge of the coverslip to draw the water under the coverslip.
9. Examine the stained slides under the compound microscope. Start using low power to focus on an area of root tip cells and then examine individual cells under ×40 power.

Name ________________________________ Date ______________

Conclusions

1. List and briefly describe each stage of mitosis.

2. Sketch a cell at each stage of mitosis. Be sure to label your sketches with the stage.

3. Does it seem that there are more cells in some stages of mitosis than others? Explain why this might occur.

4. In what stage(s) of mitosis is it easiest to see the chromosomes? Why is it easier to see the chromosomes in these stages?

5. Why did we use root tips rather than other areas of the plant to complete this lab?

Name________________________________ Date______________

CHAPTER 6
Producing Genetically Modified Organisms

LAB ACTIVITY 6.1
Cloning the Old-Fashioned Way: Asexual Propagation of Plants

Purpose

In this activity, you will utilize different methods of asexual reproduction to produce exact genetic copies of plants, also known as a clone. The relatively simple process of cloning plants by asexual propagation ensures that desirable genetic traits will be passed on from the parent plant to a daughter plant.

Materials

- Large mature geranium plant
- One potato (for every four students)
- Garlic bulb
- Large *Rex begonia* plant
- Knife
- Pins
- Toothpicks
- Rooting medium (1:1 mix of peat moss and perlite)
- Plastic flowerpots
- Rooting hormone
- Plastic wrap
- Tape for labeling

Procedure A: Cloning by Separation

1. Separation is a method of asexual reproduction in which parts of a plant are separated from the parent plant and forced to grow into a whole new plant that is an exact genetic copy of its parent. Many plants that produce bulbs and corms, like daffodils, tulips, lilies, onions, amaryllis, garlic, gladiolas, and crocuses, can be cloned using this technique.
2. Take your garlic bulb and gently break it apart into its individual cloves. Garlic is actually a corm, which is a type of underground stem. Each clove of garlic has the potential to grow into a mature garlic plant that is a clone of its parent.
3. Fill a plastic pot with rooting medium and plant one clove of garlic, pointy side up, into the medium. Cover the clove with rooting medium and water it slightly.

4. Label the pot with your name and place it near a sunny window or under a grow light.
5. Observe the growth of your garlic clone over the next few weeks.

Procedure B: Cloning by Division

1. Division is another method of asexual reproduction in which parts of a plant are divided from the parent plant and forced to grow into a whole new plant that is an exact genetic copy of its parent. Take your potato and carefully cut it into four pieces.
2. Fill a plastic pot with rooting medium and place one of the pieces of the potato in the medium. Cover it with rooting medium and water it just enough to moisten the medium. Label the pot with your name and place it near a sunny window or under a grow light.
3. Observe the growth of your potato clone over the next few weeks.

Procedure C: Cloning by Stem Cutting

1. Stem cutting is another method of asexual reproduction in which a portion of a plant stem is cut from the parent plant and forced to grow into a whole new plant that is an exact genetic copy of its parent.
2. Fill a plastic pot with rooting medium. Water the medium just enough to make it moist.
3. From the geranium plant, select a stem that is approximately 3–5 inches long and contains at least two nodes. Using the knife, carefully cut the stem from the parent plant at a 45° angle just below a node.
4. Gently strip away any leaves from the lower portion of the stem and also remove all flowers.
5. Carefully dip the cut end of the stem into the rooting hormone and tap off any excess.
6. Carefully insert the cut end into the rooting medium.
7. Label the pot with your name and place it near a sunny window or under a grow light.
8. Observe the growth of your geranium clone over the next few weeks.

Procedure D: Cloning by Leaf Vein Cutting

1. Leaf vein cutting is another method of asexual reproduction in which the leaf is cut from the parent plant and forced to grow into a whole new plant that is an exact genetic copy of its parent.
2. Select a healthy leaf from the *Rex begonia* plant. Using the knife, cut the leaf from the plant about halfway down the petiole.
3. Fill a plastic pot with rooting medium and water it slightly to make it moist.

4. Carefully use a knife to make four to six small cuts along different veins on the leaf. Use a toothpick to gently dust a very small amount of rooting hormone onto each cut.
5. Carefully place the leaf right side up and press it into the rooting medium. Hold the leaf down on the soil by using three or four pins to keep it in place.
6. Insert three or four toothpicks into the soil near the edges of the pot and cover the pot with plastic wrap, making a small tent. This will help to keep the leaf moist and increase the chances of a successful leaf clone.
7. Place your leaf near a sunny window and observe the development of your leaf-cutting clone over the next few weeks.

Name ______________________________ Date ______________

Conclusions

1. Which method of asexual reproduction produced a clone in the least amount of time?

2. Which method of asexual reproduction took the longest to produce a clone?

3. What are five methods of producing plant clones by asexual reproduction?

4. Why is the asexual reproduction of plants considered cloning?

5. What are two advantages of asexual reproduction?

Name ______________________________ Date ______________

6. What is a disadvantage of asexual propagation?

7. What are five plant parts that are commonly used for cloning?

8. What type of plant hormone did you use in this experiment (be specific), and why was it used?

9. Did all of the cloning methods prove to be successful? If not, which ones did not work, and why?

Name____ Date ______________

LAB ACTIVITY 6.2
Aseptic Technique

Purpose

The purpose of this lab is to learn the basic rules and practices of aseptic technique, the technique used to handle bacteria and other sensitive laboratory material.

Materials

- Gloves
- Powder or liquid that glows under UV light
- Handheld UV light
- Nutrient agar
- Petri dishes
- Sterile loops

Procedure

Part I

1. Follow the directions of your teacher to complete Part I of this laboratory activity.

Part II

1. Each student should put on a pair of laboratory gloves.
2. Students should obtain some powder and rub it around on their gloves, being careful not to get any powder on their skin.
3. The teacher should show students the proper way to remove their gloves without contaminating their hands.
4. Students should attempt to remove their gloves without getting any of the powder on their skin.
5. Check students' skin with the UV light to determine how successful they were at removing their gloves without contaminating their skin with the powder.

Name ______________________ Date ______________

Part III

1. Each student should put on a pair of laboratory gloves.
2. Students should obtain some powder and rub it around on their gloves.
3. Each student should get a Petri dish with some nutrient agar and a sterile loop.
4. The teacher should show students the proper way to streak a Petri dish with the sterile loop.
5. Students should practice streaking their plates with their sterile loops.
6. Use the UV light to determine if the students have contaminated their Petri dishes with the powder.

Name ________________________________ Date ______________

Conclusions

1. What is sterile technique? Why is it important to practice sterile technique when handling bacteria?

   ____

   ____

   ____

   ____

2. Why is it important for scientists and doctors to be able to remove their gloves without touching their own skin?

   ____

   ____

   ____

   ____

3. Why do pathogens (microorganisms that cause disease) spread more quickly in areas with high populations of humans?

   ____

   ____

   ____

   ____

4. What can you do in your laboratory setting to prevent contamination of your laboratory materials?

   ____

   ____

   ____

   ____

Name____ Date_______________

LAB ACTIVITY 6.3
Pest-Resistant Corn

Purpose

Students will demonstrate the effectiveness of genetically modified pest-resistant corn.

Materials

- Bt corn seed
- Wild-type (non-Bt) corn seed
- Four-cell plant pots
- Potting soil
- Large plastic tub with lid
- Fluorescent light bar or grow light
- Moths or caterpillars

Procedure

1. Students will plant Bt corn seed in one set of four-cell pots and wild-type corn in another set of four-cell pots for a total of eight plants.
2. Students will grow the corn indoors under artificial lighting (or outdoors if weather permits).
3. When corn plants are 6–8 inches tall, students will place the plants inside the plastic tub with the grow light suspended above the plants.
4. Students will place several moths inside the plastic tub with the corn.
5. Students will observe the results over several days.
6. Moths or caterpillars can be caught from the wild or purchased from scientific supply companies. Moths can be easily caught by placing an open cardboard box under a light outside overnight. In the morning, look in crevices of the box to find moths.

Name ______________________________ Date ______________

Conclusions

1. What happened to the wild-type corn? The Bt corn?

2. What do you think caused the results observed in question 1?

3. What would be the advantage to planting Bt corn over wild-type corn?

4. Can you think of any disadvantages to using Bt corn rather than wild-type corn?

Name____ Date______________

CHAPTER 7
Animal Cloning

LAB ACTIVITY 7.1
Selective Breeding: Salt Tolerance

Purpose

In this activity, you will investigate selective breeding and how it is used to identify desirable genetic traits, in this case salt tolerance in plants. Salt tolerance is the ability for certain species of plants to grow in soil with increased saline content. These types of plants are known as halophytes ("salt plants"). The high salt levels in some soils used for agriculture have had a negative impact on crop productivity. Increasing the ability of certain plants to grow in these conditions may help to boost the productivity of crops. Some plant scientists believe that there will be a time when some food crops will be genetically engineered to grow in salt water!

Materials

- Table salt
- Water
- Balance or scale (accurate to the nearest tenth of a gram)
- Salt-tolerant plants (hosta, festuca grass, sedum, honeysuckle, larch, red oak, paper birch)
- 500-ml beaker
- Barley seeds
- Two 2-liter plastic soda bottles with caps
- Utility knife
- Two 5-inch-long candlewicks or thick cotton strings
- Growth medium (1:1 ratio of perlite and peat moss)
- Hydroponics solution (1/2 teaspoon of Peters 20-20-20 plant fertilizer per 500 ml water)
- Tape for labeling

Procedure A: Salt Tolerance of Barley

1. Barley is a grain crop that has shown moderate salt tolerance. In this experiment, you will identify the salinity levels in which different varieties of barley can be grown. First, you must construct a plant growth test chamber. This will be done by modifying a 2-liter plastic soda bottle. The growth chamber used in this experiment has been adapted from the University of Wisconsin's Bottle Biology program.
2. Using the utility knife, carefully cut off the top of the soda bottle, approximately 5 inches from the bottom of the bottle.
3. Next, with help from your instructor, bore a small hole in the top of the cap of the soda bottle.

Name____ Date________________

4. Now, stick the candlewick or thick cotton string through the hole in the cap and screw it onto the bottle top. Approximately half the wick should be through the hole.
5. Mix up 500 ml of your hydroponics solution by adding 1/2 teaspoon of dry plant fertilizer with 500 ml of water.
6. Pour approximately 400 ml of hydroponics solution into the bottom portion of the soda bottle. This will act as the nutrient reservoir for your plants. Invert the top of the soda bottle into the bottom reservoir.
7. Add approximately 400 ml of growth medium into the top portion of the bottle, making sure that the cotton wick is suspended in the middle of the soil.
8. Finally, cover the reservoir portion of your plant growth chamber with a piece of paper and tape it into place. Make sure the top of the paper is aligned with the top of the water level. This will help to reduce the growth of algae in your nutrient solution. Your plant growth chamber is now complete and ready for use. It should resemble the one in Lab Figure 7–1.
9. Next, plant 10 barley seeds in the growth medium. Label your growth chamber with your name and the specific variety of barley you are testing. Once the barley sprouts and begins to grow, you will slowly begin to increase the salinity of the nutrient solution over the next few weeks to determine the plants' salt tolerance.
10. Complete the missing data in Lab Table 7–1, which you will use to calibrate the salinity of your nutrient solution. Round each numerical value to the nearest tenth.

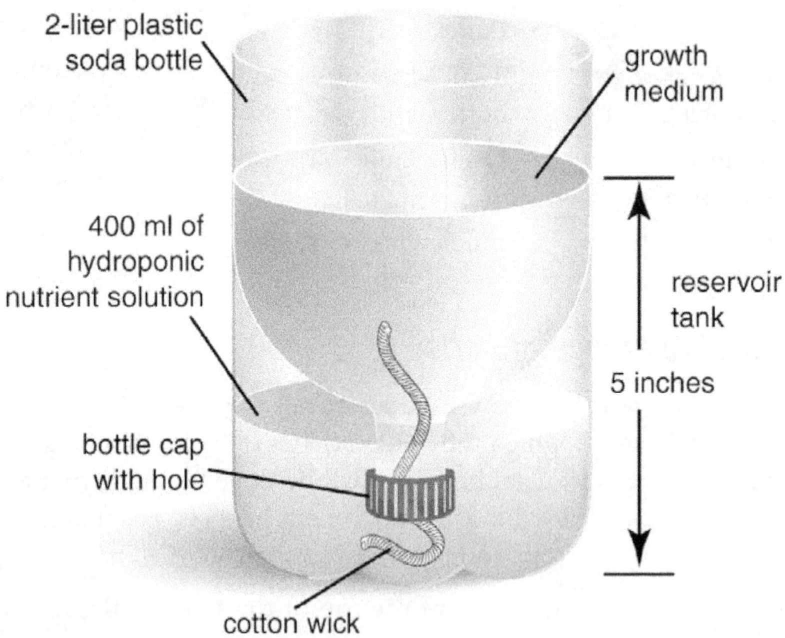

Lab Figure 7–1
Plant Growth Test Chamber

Name______________________________ Date______________

11. After the barley has been growing for a few days, record the required data in Lab Figure 7–2. Then carefully lift out the growth chamber from the reservoir and empty the old nutrient solution out of the reservoir. Mix up a new batch of nutrient solution (see step 5) and pour 400 ml of fresh solution into the reservoir. Now increase the salinity of the nutrient solution to 3 parts per thousand. This will be done by adding the amount of salt you determined from your calculations in Lab Table 7–1 to the nutrient solution in your reservoir. Replace the growth chamber and observe the growth of the plants over the next few days. Record the required data in the format shown in Lab Figure 7–2.
12. Increase the salinity level to 5 parts per thousand, using the procedure as explained in step 11. Observe the plants and record your data using the format shown in Lab Figure 7–2. Continue increasing the salinity level as shown in Lab Table 7–1 until the plants die or until you have reached the salinity level of seawater (35 parts per thousand).

LAB TABLE 7–1 Salinity Values

100% seawater	35 ppt	3.5 g/100 ml	14 g/400 ml
50% seawater	18 ppt		
	9 ppt		
	5 ppt		
	3 ppt		

Procedure B: Salt-Tolerant Perennial Screening

1. The use of salt on roads and sidewalks in the colder regions of the United States to melt ice during winter has had negative effects on many landscape plants. Identifying perennial plant varieties that are salt tolerant may prevent millions of dollars of damage done to plants each year by road salt. In this experiment, you will determine the salt tolerance of a variety of popular landscape plants.
2. Begin this experiment by constructing a plant growth test chamber like the one in Procedure A. You will also have to mix up a batch of nutrient solution for your reservoir.
3. Next, choose a perennial plant provided by your instructor to screen. Carefully transplant the plant into your growth chamber. Depending on the type of plant you are testing, you might have to shorten the top portion of the growth chamber so that the top of the plant fits.
4. Observe the growth of the plant over the next few days and record your data in the format shown in Lab Figure 7–2. Use the same procedure for increasing the salinity level every few days that you used in Procedure A. Continue to monitor the growth of your plant until it dies or until you have increased the salinity to 35 parts per thousand.

Name____ Date________________

Plant Type	Salinity Level, Plant Appearance, and Number of Days Exposed				

Lab Figure 7–2
Salt Tolerance Screening

Name ____ Date ____________________

Conclusions

1. What was the highest salinity level that your particular variety of barley survived in? What percentage of seawater did this represent?

____

____

____

2. Using data from your classmates, what was the highest salinity level achieved in this experiment? What percentage of seawater does this represent?

____

____

____

3. What is the significance to biotechnology in running experiments like this?

____

____

____

4. Explain how genetic engineering might benefit from this type of experiment.

____

____

____

5. What type of perennial plant did you test, and what was the highest salinity level that it survived in?

____

____

____

6. Using data from your classmates, what type of perennial plant grew in the highest salinity level? What was the level of the salinity that it grew in?

____

____

____

Name____ Date______________

LAB ACTIVITY 7.2
Modeling DNA Structure

Purpose

The purpose of this activity is to model the structure of a DNA molecule.

Materials

- Pipe cleaners (two different colors)
- Beads (four different colors)
- String or thread
- Slips of paper and tape or small sticky notes

Procedure

1. Use the following materials to build a model of a DNA molecule:
   - Pipe cleaner color 1 represents the deoxyribose sugar
   - Pipe cleaner color 2 represents the phosphorus group
   - Bead color 1 represents adenine
   - Bead color 2 represents thymine
   - Bead color 3 represents guanine
   - Bead color 4 represents cytosine
   - String or thread represents the hydrogen bonds that connect the complementary bases
2. Using the slips of paper and tape or small sticky notes, label all parts of the DNA model. Include the following parts:
   - Deoxyribose sugar
   - Phosphorus group
   - Adenine
   - Thymine
   - Cytosine
   - Guanine
   - Hydrogen bonds
   - Covalent bonds
3. Optional Extension Activity: Students create edible models using food products as the parts of the DNA molecule. Each student or group of students can create their own unique edible DNA model by determining which foods to use for each portion of the DNA molecule.

Name ______________________________ Date ______________

Conclusions

1. DNA is composed of units called nucleotides. What are the components of a nucleotide?

2. The percentage of nucleotides in a particular DNA molecule containing thymine is 35 percent. What percentage of nucleotides in this DNA molecule will contain adenine? Cytosine? Guanine?

3. Scientists often use models. Why might it be useful to use a model when learning about a substance like DNA?

4. Explain how the structure of DNA can determine the physical traits of an adult organism.

Name______________________________ Date______________

CHAPTER 8
Plant Cloning

LAB ACTIVITY 8.1
Herbicide-Resistant Transgenic Plants

Purpose

In this activity, you will investigate how genetic engineering has been used to create plants that are resistant to a specific herbicide. The negative impact that weeds have on agricultural productivity has been an age-old problem for farmers. The use of selective and nonselective herbicides has helped to control weeds but can be costly, toxic, and sometimes harmful to the environment. The development of a plant that is resistant to Monsanto's Roundup, a nonselective herbicide that kills a wide variety of weeds, is relatively inexpensive, and environmentally safe, and one of the most exciting uses of biotechnology and genetic engineering to date.

Materials

- Potting soil
- Large flowerpots
- Roundup Ready seeds (canola, corn, cotton, soybeans)
- Weed seeds
- Traditional seeds (canola, corn, cotton, soybeans)
- Roundup herbicide
- Safety goggles
- Rubber gloves
- Apron
- Tape for labeling
- Graph paper or spreadsheet computer application
- Colored pencils

Caution: Before applying the Roundup herbicide, follow all of the directions as outlined by your instructor about the safe handling of pesticides. Even though Roundup is a very safe product, you should follow all directions carefully. Wear your protective goggles, gloves, and apron when handling the herbicide. Apply the herbicide to your plants only when outdoors and at a safe distance away from the school building. After the herbicide has been applied, do not touch your plants or pots, and be sure to leave them outside until the herbicide has completely dried. Once the herbicide is dry, it is safe to return the plants to your classroom. Handle the plants only while wearing your protective gloves at all times.

Procedure A

1. Fill two flowerpots with soil. Label one pot "Roundup Ready herbicide resistant" and the other pot "no herbicide resistance."

2. Obtain the weed seeds from your instructor and plant them in the soil of both flowerpots.
3. Obtain the Roundup Ready seeds from your instructor and plant them in the pot labeled "Roundup Ready herbicide resistant."
4. Obtain the traditional crop seeds from your instructor and plant them in the pot labeled "no herbicide resistant."
5. After you have planted all of your seeds, label each pot with your name and place them near a sunny window or under a grow light. Carefully water your seeds and observe them each day for signs of germination.
6. Once both your weeds and crop seeds have sprouted in both pots, wait until the plants have grown their first true leaves before applying the first application of Roundup herbicide.
7. Put on your goggles, gloves, and apron and take both of your pots to an area designated by your instructor a safe distance away from the school building. Using the directions from your instructor, carefully apply the first application of Roundup to all of the plants in both pots. Leave the pots outside until the Roundup has dried. You may then return the plants to the classroom.
8. Observe the growth of all the plants in both pots over the next few days. Your instructor may decide to have you apply another application of Roundup, depending on the results of your first application.

Procedure B

1. Use the data on the acreage of transgenic crops shown in Lab Table 8–1 to construct a multiple line graph that illustrates the specific type of traits used in transgenic crops planted between 1996 and 2002. The *x*-axis should be labeled "year" and the *y*-axis "millions of acres." Use a different-colored pencil for each line and create a key for your graph.

LAB TABLE 8–1 Transgenic Crop Traits Used Between 1996 and 2002 (millions of acres)

	1996	1997	1998	1999	2000	2001	2002
Herbicide tolerance	1.5	17.0	46.5	70.5	82.0	101.5	110.5
Insect tolerance	2.8	12.3	16.8	22.0	20.8	19.5	25.3
Bt/herbicide tolerance	0.0	0.1	6.3	7.3	7.3	10.5	11.0

Name ______________________________ Date ______________

Conclusions

1. Describe the appearance of both the crop and weed plants growing in both pots before applying the Roundup herbicide.

____

____

____

____

2. Describe the appearance of both the crop and weed plants in both pots after applying the Roundup herbicide.

____

____

____

____

3. Did the results of your experiment support the benefits of using genetic engineering to develop herbicide-tolerant plants? Explain.

____

____

____

____

4. Describe in detail the exact way that Roundup Ready transgenic plants are able to tolerate exposure to the Roundup herbicide.

____

____

____

____

5. List at least 10 different types of crops that have been or are being developed to be herbicide resistant.

____

____

____

____

Name______________________________ Date__________

6. Using the graph you created in Procedure B, what are the three types of traits that were used in transgenic crops between 1996 and 2002?

____

____

____

____

7. Using the graph you created in Procedure B, identify which type of transgenic trait is used most widely in crops planted in 2002?

____

____

____

____

8. Using the data from Lab Table 8–1, calculate the percentage increase of herbicide-tolerant crops planted between 1996 and 2002.

____

____

____

____

9. Did the results of the experiment you conducted in Procedure A explain the increase in herbicide-tolerant crop use over the 7-year period? Why or why not?

____

____

____

____

Name____ Date____________

LAB ACTIVITY 8.2
Grafting

Purpose

The purpose of this lab is to introduce the principles of grafting, a traditional method of asexual reproduction for woody stemmed plants.

Materials

- Two varieties of woody stemmed plants (hibiscus work well)
- Grafting knives
- Grafting rubbers (or rubber bands, cut open to make straight pieces of rubber)
- Laboratory film (or plastic wrap)
- Distilled water
- Spray bottle

Procedure

1. Choose two varieties of woody stemmed plants. Hibiscus plants work well because they grow quickly, are small enough to have in a classroom environment, and come in a wide variety of colors.
2. One plant will be the stock plant, and the other will be the donor plant that produces the scion. The stock plant will be the main plant that remains rooted, and the scion will be grafted onto the stock plant.
3. Use the grafting knife to cut a piece off the scion. *Be very careful with the grafting knife. It is very sharp!*
   a. To perform a cleft graft:
      i. Cut a branch off the donor plant—this is the scion. Be sure the branch has a bud on the end. Remove all leaves from the scion. *Do not remove the bud.*
      ii. Cut the base of the scion into a point by using the grafting knife to cut each side of the base at an angle.
      iii. Cut a branch off the stock plant. Be sure the stock branch is approximately the same diameter as the scion. Discard the branch.
      iv. Using the grafting knife, cut a slice across the diameter of the base of the branch left on the stock plant.
      v. Insert the point of the scion into the slice on the stock plant.

vi. Be sure to line up the cambium layer (vascular tissue/meristematic tissue) of the stock plant with that of the scion.

b. To perform a bud graft:

i. Find a bud protruding from the stem of the donor plant. Use the grafting knife to slice the bud off the stem of the donor plant. Be sure to include some cambium layer and bark on either end of the bud in your cut. This is the scion.

ii. Cut a matching area from the stock plant. Be sure to make the cut on a main stem of the stock plant.

iii. Place the scion in the hole on the stem of the stock plant.

iv. Be sure to line up the cambium layer (vascular tissue/meristematic tissue) of the stock plant with that of the scion.

4. Hold the scion firmly in place and wrap with the grafting rubber or with a rubber band that has been cut open to hold the scion in place. Once you have firmly wrapped the grafting rubber around the scion, slip the free end of the grafting rubber underneath to "tie off."

5. Carefully wrap the entire area with laboratory film or plastic wrap. Be sure the area where the scion meets the stock is completely covered by the laboratory film. If the scion dries out or becomes contaminated with bacteria or fungi, the graft will not take.

6. Place the stock plant in a location with adequate sunlight and humidity. It is important that the grafted area is not dried out.

7. The donor plant may be used to produce additional scions or as another stock plant.

8. Check on the graft approximately once per week by removing the laboratory film and grafting rubber. Ensure that the area is still moist and that there is no fungal growth. If necessary, spray the grafted area lightly with distilled water. Rewrap with a clean grafting rubber and clean laboratory film.

9. Continue checking on the graft until it has "taken" and the outside of the stock plant (bark) has grown around the grafted area.

Name ______________________________ Date ______________

Conclusions

1. What are some advantages of grafting one variety of a woody plant to another?

____

____

____

____

____

2. Plants must be closely related to each other in order for grafting to occur easily. Why do you think this is the case?

____

____

____

____

____

3. List two circumstances that would cause a graft to fail.

____

____

____

____

____

4. Why is it so important to line up the cambium layers of the stock and the scion?

____

____

____

____

____

Name____ Date__________________

LAB ACTIVITY 8.3
Micropropagation of Plants

Purpose

The purpose of this lab is to introduce students to the practice of tissue culture, or micropropagation, of plants as a means of rapid asexual reproduction.

Materials

- Stock plants (African violets work well) (cultures may also be purchased from science supply stores)
- Multiplication media (can be purchased from a science supply store)
- Rubbing alcohol
- Forceps
- Culture tubes with tops
- Culture tube rack
- Scalpel
- Dissection needle
- Gloves
- Laboratory film
- Petri dish (optional)
- Lab hood (optional)
- Grow light (optional)

Procedure

1. ***Tip:*** Always use aseptic technique when handling media, Petri dishes, and plant material to avoid contamination of your experiment.
2. Prepare the culture tubes by filling them 2/3 full with melted multiplication media. Allow the media to solidify. You may want to place the culture tubes at an angle while the media solidifies to create a sloping surface (optional). Cover the culture tubes to avoid contamination.
3. The parent tissue used to create new clones, or plantlets, is called the explant. Using the scalpel, cut a small piece of leaf tissue explant from the stock plant.
4. Perform steps 5–7 under a laboratory hood if available.
5. Place the explant in a Petri dish or other small container and use a small amount of rubbing alcohol to sterilize the surface of the explant. Remove any excess rubbing alcohol from the bottom of the dish.
6. Use the scalpel and the forceps to cut the explant into very small pieces (1–3 mm in diameter).
7. Use the forceps to place two to three pieces of explant on the surface of the multiplication media in each culture tube. Recap the culture tubes and use laboratory film to seal the caps.
8. Place the sealed culture tubes on a rack under a grow light, if available.
9. Observe your culture tubes daily for signs of growth.

Name ______________________________ Date ____________

Conclusions

1. What are the advantages of using micropropagation (tissue culture) rather than traditional methods of vegetative propagation?

2. Why is it important to use sterile technique when performing micropropagation procedures?

3. Why do you think micropropagation is sometimes referred to as "plant cloning?"

4. What properties of plants allow the explant to develop into multiple plantlets?

Name____ Date____________

CHAPTER 9
Biotechnology in Plant Science

LAB ACTIVITY 9.1
Penicillin: An Accident Waiting to Happen

Purpose

In this activity, you will recreate one of the most amazing accidental scientific discoveries ever made. In 1928, British bacteriologist Alexander Fleming accidentally left a bacterial culture unattended. Eventually, he found that the culture had been contaminated by the growth of a fungus. To his astonishment, he observed that the growth of the fungus had killed off the strain of bacteria. After further investigation, Fleming discovered that the fungus *Penicillium* was secreting a chemical that was deadly to bacteria. He called the chemical penicillin, which became the first antibacterial drug used to treat infections. Today, advances in biotechnology have led to purified penicillin and other pharmaceutical agents produced by microorganisms that have had a profound effect on human health.

Materials

- Three sterile Petri dishes containing nutrient agar, with lids
- Two sterile swab applicators
- Living fungi culture of high-yield *Penicillium notatum*
- Living bacterial culture of *Staphylococcus epidermidis*
- Plastic tape
- Rubber gloves
- Safety goggles
- Biohazard autoclave bag
- Colored pencils

Caution: Follow all of the directions as outlined by your instructor about the safe handling of the bacterial and fungal cultures used in this experiment.

Procedure

1. Obtain three sterile Petri dishes from your instructor. Do not open them. Keep the lids on the dishes at all times. Mark the bottom of one of the Petri dishes with an "A," one with a "B," and the third with a "C." Also mark each dish with your name. Petri dish A will contain only the *Staphylococcus epidermidis* culture. Petri dish B will contain both the *Staphylococcus epidermidis* culture and the *Penicillium notatum* culture. Petri dish C will contain only the *Penicillium notatum* culture. Enter this information in the format shown in Lab Figure 9–1.

Name____ Date____________________

	Microorganisms Present	Observations
Petri Dish A		
Petri Dish B		
Petri Dish C		

Lab Figure 9–1

2. Put on your gloves and safety goggles. Using one of your sterile applicator swabs, obtain a sample of the *Staphylococcus epidermidis* culture from your instructor. Quickly open Petri dish A and inoculate the nutrient agar by gently rubbing the swab containing the *Staphylococcus epidermidis* culture on the growth medium, using the technique shown by your instructor. Quickly replace the lid of the dish.
3. Using the same swab that contains the *Staphylococcus epidermidis* culture, quickly open Petri dish B and inoculate the nutrient agar by gently rubbing the swab on the growth medium, using the technique shown by your instructor. Quickly replace the lid of the dish.
4. Discard the swab that contains the *Staphylococcus epidermidis* culture in the biohazard autoclave bag provided by your instructor.
5. Using a new sterile applicator swab, obtain a sample of the *Penicillium notatum* culture from your instructor. Quickly open Petri dish C and inoculate the nutrient agar by gently rubbing the swab containing the *Penicillium notatum* culture on the growth medium, using the technique shown by your instructor. Quickly replace the lid of the dish.
6. Using the same swab that contains the *Penicillium notatum* culture, quickly open Petri dish B. Inoculate the nutrient agar by gently rubbing the swab containing the *Penicillium notatum* culture on the growth medium, using the technique shown by your instructor. Quickly replace the lid of the dish.
7. Discard the swab that contains the *Penicillium notatum* culture in the biohazard autoclave bag provided by your instructor.
8. Seal the sides of each dish with plastic tape and place them in a warm, dark area designated by your instructor.
9. Observe the growth of the cultures in each plate. When the growth of the two cultures is complete, record your observations in the format shown in Lab Figure 9–2.
10. Next, within a circle, draw the growth of the cultures observed in each dish. Label each circle. Use a different-colored pencil to represent each culture (*Staphylococcus epidermidis* and *Penicillium notatum*) and create a key.

Name____________________________________ Date______________

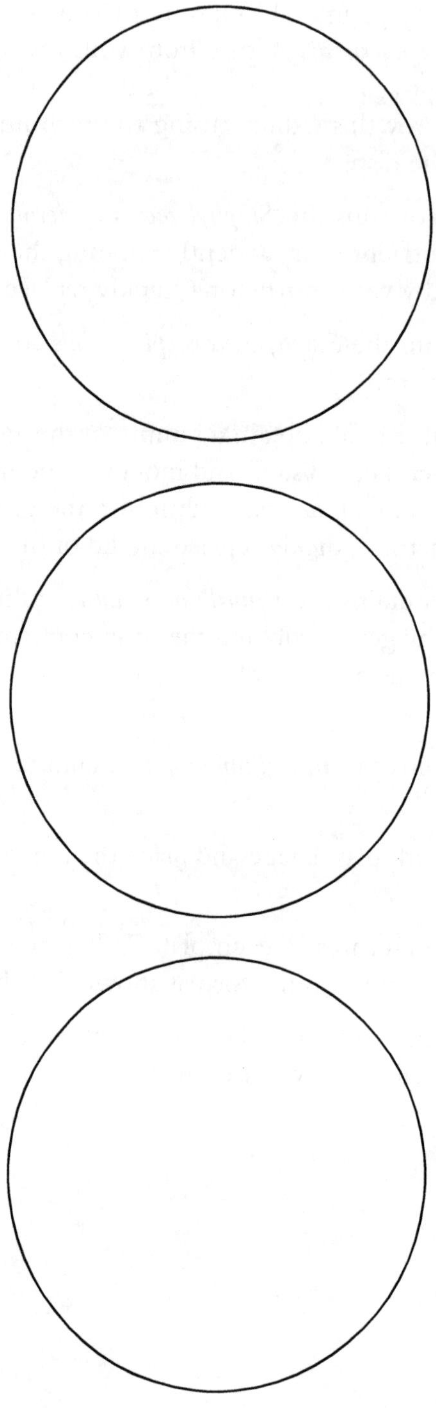

Lab Figure 9–2
Growth of Observed Cultures

Name ____ Date ______________

Conclusions

1. Why was it important to grow cultures of the *Staphylococcus epidermidis* and *Penicillium notatum* in two separate Petri dishes?

____

____

____

____

____

2. In detail, describe the results of your experiment.

____

____

____

____

____

3. Explain how genetic engineers might benefit from this experiment.

____

____

____

____

____

4. Describe other experiments that you would need to conduct to prove that penicillin could be used as an effective antibiotic.

____

____

____

____

____

Name ________________________________ Date ____________

CHAPTER 10
Biotechnology in Animal Reproduction

LAB ACTIVITY 10.1
Analysis of Cryopreserved Bull Semen

Purpose

Students will calculate the number of bull sperm per unit of volume.

Materials

- Frozen straws of bull semen
- Microscope
- Hemacytometers with coverslips
- Pipettor with tips
- 3 percent saline solution

Procedure

1. Students will properly thaw a frozen straw of bull semen using a water bath at 37°C.
2. Students will make a 1:10 dilution of thawed semen in 3 percent saline.
3. Students will thoroughly mix semen and the diluent by inverting the tube several times
4. Students will position a coverslip squarely over the counting chamber of the hemacytometer and place the hemacytometer under the microscope.
5. Students will load sample on the hemacytometer by placing semen in the angle of the coverslip and grid surface using a pipettor and watch the diluted semen flow under the coverslip until the chamber is full.
6. Students will locate the counting area of the hemacytometer under low power and then switch to high power for counting. Students may need to allow the sperm cells to settle for 2–3 minutes before beginning to count.
7. Students will count the five large squares (80 small squares) diagonally from the upper left to the lower right of the triple-lined area of the hemacytometer. Be consistent in how you count the cells that are touching a line. For example, count only those cells that are touching the upper and left lines, not those touching the lower and right lines. This will eliminate duplication in counting.

8. Students will calculate the sperm cell concentration as follows:
   - Semen dilution factor is 10. (Remember the original dilution was 1:10.)
   - Five large squares (80 small squares) were counted on the hemacytometer.

   Sperm concentration/ml = (dilution factor)(count in five squares)(50,000)

For a further explanation of these calculations, see the University of Wisconsin Animal Science website at http://www.ansci.wisc.edu/jjp1/ansci_repro/lab/procedures/hemacytometer/Hemocytometer%20use.html.

Name______________________________ Date______________

Conclusions

1. What is the value in knowing the sperm concentration in a sample of semen?

2. Describe the appearance of the sperm in the counting area.

3. Why was the sperm diluted in 3 percent saline?

4. Why is it necessary to know the volume of the hemacytometer counting chamber?

Name____ Date________________

LAB ACTIVITY 10.2
Feed Preservation: Silage

Purpose

The purpose of this experiment is to build and test a silo model for preserving grass as silage, which is used as feed for cows, through the process of silage fermentation. In this activity, you will build your silo using a 2-liter soda bottle and monitor the silage over a 4-week period. During the 4-week observation period, you will identify the three main stages of silage fermentation in your silo. These are the respiration stage, the fermentation (or pickling) phase, and the stable phase. Silage that develops a pH of 4.2 or lower is considered stable and may be kept for years as long as air is excluded.

Materials

- 2-liter soda bottle with cap
- Utility knife
- Freshly cut grass
- Scissors
- Tape
- Petroleum jelly
- Tongue depressor
- 16-inch section of thin plastic tubing
- Two small test tubes
- Bromthymol blue pH solution
- pH test papers (one with range reaching 7.0 and another with range of 3.0–5.5)
- Balance or scale
- Parafilm sealing film
- Safety goggles

Safety Concern: When preparing or working with the pH indicator solution, always wear safety glasses.

Procedure

1. Using the utility knife, cut the plastic bottle just below the curved top, near the top of the label.
2. Carefully cut the bottle approximately 6 inches from the bottom. The resulting top part should then fit tightly into the bottom, with the cap near the same level as the sides. Trim either the top or the bottom to ensure a tight fit. See Lab Figure 10–1.
3. Collect enough fresh grass to fill the top piece. Make sure the cap is on the bottle and pack the grass into the top tightly.
4. Using the format shown in Lab Figure 10–2, record the color of the grass.
5. Weigh the grass using a balance or scale. Record the mass of the grass to the nearest tenth of a gram in Lab Figure 10–2. The weight of the grass should be about 100 grams.

Name ____________________ Date ____________

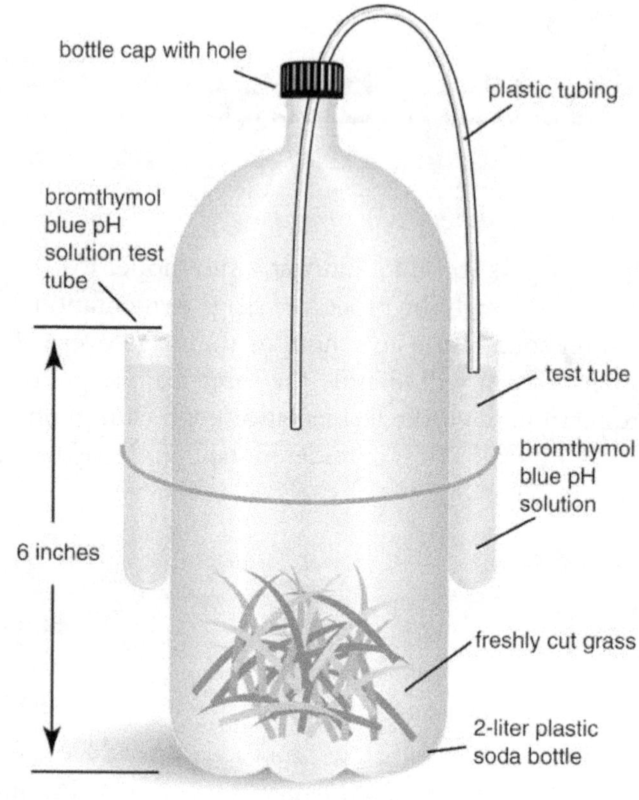

Lab Figure 10–1
Mini-Silo Model

	Color of Grass	Mass of Grass	pH of Grass	Color of Bromthymol Blue Solution
Date				
Date				
Date				

Lab Figure 10–2
Silage Fermentation Data

6. Using the pH paper, test the pH of the grass and record its pH value in Lab Figure 10–2.
7. With help from your instructor, bore a hole in the top of the bottle cap and insert approximately 2–3 inches of the plastic tubing through the hole.
8. Using the scissors, carefully chop the grass into lengths of approximately 1 inch. Gather up all the grass and tightly pack it into the top of the bottle.
9. Carefully put the top into the bottom with the cap facing up. Gently press the top into place, compressing the grass to remove as much air from within the packed grass as possible. The more tightly the grass is packed, the better your results will be. With help from a classmate, secure the top in place by tightly taping the top of the bottle to the bottom.
10. Seal the interface between the top and the bottom of the bottle with a layer of petroleum jelly applied with a tongue depressor. Make sure all gaps are filled to ensure an airtight seal. The interior of the top must be airtight for fermentation to occur.
11. Seal the area on the cap where the tube enters the bottle.
12. Attach two regular test tubes vertically to the side of the "silo" with two pieces of tape running horizontally.
13. Fill each test tube about three-quarters full with the bromthymol blue pH solution. Insert the tube at least halfway into one of the test tubes and secure it in place with tape. Close both test tubes with Parafilm. Record the color of the bromthymol blue pH solution in Lab Figure 10–2. Your silo is now complete and should look similar to the one in Lab Figure 10–1.
14. Place your silo in a dark area designated by your instructor and continue to make observations for a period of approximately 4 weeks. Observe and record any changes in the color of the bromthymol blue pH solution and the appearance of the grass. Record your observations in Lab Figure 10–2.
15. After approximately 4 weeks, open your silo and observe the appearance of the grass and any particular odors. Test the pH of the silage with a pH paper. The pH will be lower than 5.0 if fermentation has occurred. Record the information in Lab Figure 10–2.
16. Weigh the silage and record its mass in Lab Figure 10–2. Your experiment is now complete.

Name ____________________ Date ____________

Conclusions

1. Describe the process by which grass is preserved by fermentation.

2. Explain why it was necessary to attach two test tubes filled with the bromthymol blue pH solution.

3. Describe the color of the grass at the start of the experiment and then its color at the end of the experiment.

4. Describe the pH value of the grass at the start of the experiment and then the pH value at the end of the experiment.

Name ______________________________ Date ______________

5. Did the pH value you determined at the end of the experiment reach the desired level to create silage?

6. Was there a change in the mass of the grass between the start and the end of the experiment? Why or why not?

7. Explain why the creation of silage is considered to be a biotechnology.

8. Briefly explain why your experiment was or was not a success.

Name____ Date______________

LAB ACTIVITY 10.3
Where Are the Bacteria?

Purpose

The purpose of this lab is to determine what areas of the classroom are home to the most bacteria and other microorganisms. Students will learn how to culture bacteria and identify bacterial colonies.

Materials

- Petri dishes (sterilized)
- Large cotton swabs (sterilized)
- Nutrient agar
- Laboratory film
- Bleach or autoclave
- Lab incubator (optional)

Procedure

1. ***Tip:*** Always use aseptic technique when handling media, Petri dishes, and bacteria to avoid contamination of your experiment.
2. Prepare the Petri dishes by pouring ½–¾ of an inch of nutrient agar in the bottom half of the plates. Let the agar solidify.
3. Determine a location in the school or classroom where you will test for the presence of microorganisms.
4. Use the large cotton swabs to sample various surfaces within the location.
5. For each surface, gently rub the cotton swab across the surface. Use the cotton swab to streak a Petri dish by gently rubbing the swab across the surface of the agar. Turn the dish and streak again in the other direction.
6. Use a different dish for each surface you test. Be sure to label the dishes.
7. Use laboratory film to seal the dishes closed. Important: *Do not* reopen the dishes once they are sealed.
8. Place the dishes in an incubator set to 40°C for 24 hours or let the dishes sit at room temperature for 2–3 days.
9. You should be able to see bacterial colonies (and possibly fungal growth) on the agar after 1–3 days.
10. Compare the growth on the Petri dishes from the different surfaces to determine where the most microorganisms are found.

Tip: At the end of the lab, soak the Petri dishes in a 10:1 solution of water and bleach or autoclave before disposal.

Name ____ Date ________________

Conclusions

1. What area(s) did you test? What kinds of surfaces did you test?

2. Which surface had the most microorganisms?

3. Draw a picture of the dish with the most microorganism growth. Clearly show the bacterial colonies.

4. If there are microorganisms on almost every surface we come in contact with, why are not people sick more often?

Name______________________________ Date______________

CHAPTER 11
Biotechnology in Medicine

LAB ACTIVITY 11.1
Effects of Antibiotics

Purpose

The purpose of this lab is to determine how antibiotics affect bacterial populations.

Materials

- *Escherichia coli* bacteria culture
- Petri dishes
- Nutrient agar
- Sterile loops
- Forceps or tweezers
- Laboratory film
- Antibiotic disks
- Ampicillin
- *E. coli* with the bla gene for ampicillin resistance
- Permanent marking pen
- Laboratory incubator (optional)

Procedure

1. Pour about ½ inch of melted nutrient agar into the bottom of each Petri dish. Allow the agar to solidify.
2. Each lab group should have two Petri dishes with agar and two sterile loops.
3. Use a sterile loop to transfer a colony of bacteria from the bacterial culture to a Petri dish. Use the loop to spread the bacteria over the surface of the agar, starting at the edge of the plate and moving toward the center. Turn the dish 45° and repeat. Continue this process until the entire surface of the agar has been covered.
4. Repeat step 3 for the second Petri dish. Be sure to use a different sterile loop for each dish.
5. Obtain four antibiotic disks. Using the forceps, dip each disk into the ampicillin. Using a dry pair of forceps, place the disks in a circle in the middle of one of the Petri dishes.
6. Do not put any disks in the other Petri dish. The Petri dish without the disks will serve as your control.
7. Cover the Petri dishes and wrap them with laboratory film to prevent contamination.
8. Place the Petri dishes in a laboratory incubator set at 37°C for 36–48 hours.
9. After incubation, compare the bacterial growth in each dish.

Name ______________________________ Date ______________

Conclusions

1. Draw a diagram of each dish. Be sure to label the locations of the disks.

2. Based on your results, what can you conclude about the effectiveness of ampicillin when used against *E. coli*?

3. Scientists call the area around the disks "zones of inhibition." Why do you think they use this term?

4. Researchers and scientists sometimes engineer bacteria that contain a desirable trait to also be resistant to a particular antibiotic. Why would such engineering be useful?

Name____ Date______________

CHAPTER 12
Biotechnology in the Food Industry

LAB ACTIVITY 12.1
Food Preservation: Enzymes

Purpose

In this activity, you investigate the importance of using enzymes for helping to preserve food. Enzymes are protein molecules that are produced by organisms to control the rate of chemical reactions. The first activity will illustrate how enzymes can be used to increase the shelf life of food, in this case bread. In the second activity, you will observe how enzymes are used to preserve milk by turning it into cheese. Today, many different types of enzymes are used widely in biotechnology.

Materials

- One package quick rising yeast
- 1 tablespoon sugar
- 3 cups bread flour
- 1½ teaspoon salt
- 1 cup very warm water
- 2 tablespoons melted butter or margarine
- Vegetable oil
- Plastic wrap
- Baking pan or two small bread loaf pans
- Two large mixing bowls
- Wooden spoon
- Oven
- Two 250-ml beakers
- 400 ml whole milk
- Maltogenic amylase enzyme
- Rennin (chymosin) enzyme

Procedure A

1. Mix together in one of the large mixing bowls only 2 cups of flour plus the yeast, salt, and sugar.
2. Pour the melted butter or margarine and very warm water into the flour mixture. Mix together thoroughly with a wooden spoon.
3. Divide the dough mixture in half and place into two separate mixing bowls.
4. Sprinkle one of the dough mixtures with the correct amount of maltogenic amylase enzyme suggested by your instructor and mix it in thoroughly into the dough. Label the mixing bowl containing the dough with the enzyme.
5. Slowly add ½ cup of flour to both of the dough mixtures until the dough is moist but not too sticky. Knead each dough ball for about 5 minutes, then coat each dough ball with a small

amount of vegetable oil, and cover each bowl with a plastic wrap. Place each bowl in a warm place to rise for about 30 minutes.

6. After each dough ball has nearly doubled its volume, shape the dough into a bread loaf and place on a greased baking sheet or in two small bread loaf pans. Take a knife and score the top of the loaf containing the maltogenic amylase enzyme with an "S" shape. This will help you identify this loaf after it has been baked.
7. Set the loaves aside for another 20 minutes, after which the loaves will be baked in a preheated oven set at 350°F for about 30 minutes.
8. Once your loaves have been baked and allowed to cool completely, place each loaf in a plastic bag. Label each bag with your name and the date. Observe the two loaves over the period of 2 weeks.

Procedure B

1. In this activity, you will observe how enzymes are used to preserve milk as cheese. Pour 200 ml of milk into two 250-ml beakers. Label one beaker "enzymes added" and the other beaker "no enzymes."
2. In the enzymes-added beaker, sprinkle in the correct amount of the rennin.
3. Observe the properties of both beakers of milk over the next few minutes.

Name ______________________________ Date ______________

Conclusions

1. What is an enzyme?

2. Describe the chemical action that caused the bread dough to rise in Procedure A.

3. Did the maltogenic amylase enzyme have any effect on the bread loaf you made in Procedure A? If so, what was the effect?

4. Describe the difference in the characteristics of the two loaves of bread at the end of your experiment.

Name____ Date______________

5. What effect did the rennin enzyme have on the milk?

____

____

____

____

____

6. Did this lab investigation illustrate the role of enzymes in biotechnology? Why or why not?

____

____

____

____

____

Name ____________________ Date ____________

CHAPTER 13
Biotechnology in Ecology

LAB ACTIVITY 13.1
Bioremediation Part 1: Secondary Wastewater Treatment

Purpose

In this activity, you will design and build a working model of the secondary wastewater treatment process and observe how microorganisms act as a living filter to remove unwanted dissolved biodegradable organic compounds from wastewater.

The use of living things to clean up the environment is known as bioremediation and is a rapidly advancing biotechnology.

Materials

- Plastic 2-liter soda bottle
- Aquarium aerator
- Plastic aquarium tubing
- 8-inch section of ½-inch plastic tubing
- Sharp utility knife
- Duct tape
- Simulated wastewater (salt-free crackers dissolved in chlorine-free water)
- Mature yard compost or soil from an outdoor source (as a bacterial inoculant) or bacteria starter culture used for biological filtration in aquariums
- Plastic wrap
- Washed, pea-sized aquarium gravel
- Metric ruler
- Graph paper or computer spreadsheet program

Procedure

1. Collect all the required materials and bring them to your work area. Use the diagram in Lab Figure 13–1 to help you construct your secondary wastewater treatment tank. First, using the utility knife, carefully cut off the top third of the plastic soda bottle. Next, turn the top of the bottle upside down and insert it into the bottom part of the bottle. It should fit snugly and resemble a funnel.

2. Next, carefully pour enough simulated wastewater into the bottom of the bottle to reach about an inch below the bottom of the funnel.

Name____ Date____________

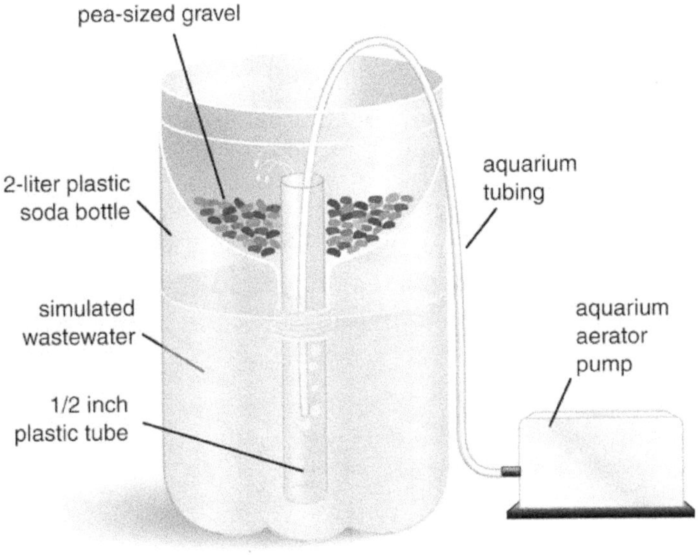

Lab Figure 13–1
Simple Secondary Wastewater Treatment Model

3. Now insert the piece of ½-inch plastic tubing through the funnel and down into the wastewater.

4. Carefully fill the funnel about half full with the gravel. Then sprinkle a small amount of soil, compost, or bacteria starter culture onto the gravel.

5. Attach the aquarium tubing to the aquarium aerator and then carefully insert the other end of the tubing into the ½-inch plastic tube that is sticking through the funnel. Plug in the aerator. You must now adjust the depth of the aquarium tube within the wastewater to create an airlift pump. When properly adjusted, the air bubbles should force the wastewater to bubble up and out of the ½-inch tube and trickle down through the gravel and back into the bottom of the bottle. It may also be necessary to alter the height of the ½-inch tubing so that the airlift pump is working efficiently.

6. Once you have adjusted the tubing to the point that the wastewater is being pumped up and onto the gravel bed, secure the tubing into position with duct tape.

7. After you have secured your tubing, place your secondary wastewater treatment model in a safe place. Cover the top with plastic wrap to prevent excess evaporation and splashing of the wastewater.

8. Now you will observe your model for a period of 1 or 2 weeks and make daily turbidity measurements. Turbidity is a measure of how clear the water is. To measure it, have one of your partners carefully lift up the funnel from your model so that you can insert a metric ruler into the bottom tank. Slowly immerse the ruler into the wastewater until you can no longer read the zero at the end of the ruler. Then record the depth to the nearest tenth of a centimeter.

This measurement will represent the turbidity. Record your daily turbidity measurements in the format shown in Lab Figure 13–2. This simple relative scale will make it possible to measure the changes in water clarity over the time of your experiment. Because some water will be lost to evaporation, it may be necessary to add clean, nonchlorinated water into your model periodically to keep it running efficiently.

9. Once you have concluded your experiment, you will construct a line graph that shows the change in the turbidity of the wastewater. The *x*-axis should be labeled "days" and the *y*-axis "turbidity in centimeters," as recorded in the format shown in Lab Figure 13–2.

Name____ Date ____________________

Date	Turbidity (centimeters)

Lab Figure 13–2

Name ______________________________ Date ______________

Conclusions

1. Describe the appearance of the simulated wastewater at the start of your experiment.

2. What purpose did the gravel serve in your model?

3. Explain the two purposes that the aquarium aerator served in your model.

4. Why was it necessary to sprinkle soil or compost on your gravel bed?

Name ______________________________ Date ______________

5. Describe the appearance of the wastewater at the end of your experiment.

6. Using your turbidity data from Lab Figure 13–2, calculate the daily rate of change in turbidity over the length of your experiment. The rate of change can be calculated by using the formula: rate of change = change in turbidity/change in time.

7. Did your wastewater treatment model successfully treat the wastewater?

8. Describe how this experiment illustrates how the process of secondary wastewater treatment classifies as a biotechnology.

Name____ Date______________

LAB ACTIVITY 13.2

Biodegradable Plastics Lab

Purpose

The purpose of this lab is to compare the rate of decomposition of biodegradable plastic with the rate of decomposition of regular plastic.

Materials

- Open-topped containers (old food storage containers work well)
- Soil
- Permanent markers
- Spray bottles
- Various types and sizes of regular plastic. Ideas include the following:
  - Bingo chips
  - Strips of zippable plastic bags
  - Strips of garbage bags
  - Styrofoam packing peanuts
- Various types and sizes of biodegradable plastic. Ideas include the following:
  - Pieces of biodegradable silverware
  - Strips of biodegradable food packaging
  - Strips of biodegradable plastic bags
  - Biodegradable packing peanuts

Procedure

1. Use water to moisten the soil until it is just slightly damp.
2. Fill the open-topped container halfway with damp soil.
3. Place similar-sized pieces of regular plastic and biodegradable plastic on the surface of the soil.
4. Use the permanent marker to make a note of where the plastic items are on the side of the container. Alternatively, draw a diagram of your container and carefully label the location of each plastic piece.
5. Record detailed observations about the appearance and size of each plastic piece.

6. Cover the plastic pieces with several inches of the damp soil.
7. Leave the container open to the air in a warm and moist location. Check on your container at least once per week and add water to the soil if it appears dry.
8. After 4–6 weeks, use your diagram or the markings on the side of the container to dig up your regular and biodegradable plastic pieces.
9. Record detailed observations about the appearance and size of each plastic piece at the end of the 4 to 6 week period.

Name ____________________ Date ____________

Conclusions

1. Based on your observations, which item(s) degraded most? Which item(s) degraded least?

2. What conditions are necessary for biodegradable items to degrade? Did your laboratory setup meet all those conditions?

3. Many biodegradable plastics are corn or soy based. What are some economic and societal implications of using corn- or soy-based plastics?

4. A majority of plastic consumer materials are still made from regular (oil-based) plastics. Describe some strategies that would encourage business and industry to use more biodegradable plastics.

5. Describe an experiment you could run as a follow-up to the results from this study.

Name____ Date________________

LAB ACTIVITY 13.3
Biogas Generators

Purpose

The purpose of this activity is for you to observe another fermentation process that is the result of naturally occurring bacteria. Methane gas, a hydrocarbon fuel commonly called natural gas, is produced by bacteria living in anoxic (oxygen-free) conditions. Researchers have utilized this fermentation process to create natural gas in a device called a biogas generator. The term "biogas" is derived from the words "biologically produced gas." In this activity, you will construct a simple biogas generator to observe the action of bacteria and to learn how biogas is produced.

Materials

- One 12-ounce soda or juice container
- 6 ounces cow manure
- Large balloon
- Rubber band
- 1 tablespoon corn syrup
- One 32-ounce container (coffee cans or plastic food containers work well)
- Measuring cup
- Measuring spoon
- 6 ounces dechlorinated water

Caution: Although the methane gas produced in the balloon contains a high water content, methane gas is flammable so it is essential to keep the balloon away from any open flame or heat source.

Procedure

1. Collect all the required materials and bring them to your work area. Mix approximately 6 ounces of manure with 6 ounces of dechlorinated water (1:1 mixing ratio) and 1 tablespoon of corn syrup in the 32-ounce container. (It is important to use water that has no chlorine in it for this experiment to be successful. Spring water or distilled water can be purchased at a grocery store, or tap water can be dechlorinated by using a dechlorinating agent available at an aquarium supply store.)
2. Carefully pour the mixture into the soda bottle so as to completely fill it. Then cover the opening at the top of the bottle with the balloon. Make sure the balloon is brought all the way down around the opening to ensure an airtight fit. Reinforce the seal between the balloon and the bottle by tightly wrapping a rubber band around the bottleneck and the balloon.

3. Label your container with your group name and place it in a well-ventilated, dark, warm place designated by your instructor. A greenhouse or outside location works well so that the smell of the manure does not disturb your classroom.
4. You will observe your biogas generator over a period of a week or two. At first, carbon dioxide gas will be produced by aerobic (oxygen-loving) bacteria. This can be released from the balloon by temporarily breaking its seal. Eventually, anaerobic (no-oxygen) bacteria, called methanobacteria, will begin to digest the manure and produce methane gas, which will begin to collect in the balloon. If the balloon becomes filled, release the gas by lifting the side of the balloon away from the opening of the bottle. The balloon should refill with gas in a few days.

Name ____________________ Date ____________

Conclusions

1. Explain why it is necessary to use water that has no chlorine in it for this experiment.

2. Why was it necessary to mix corn syrup in the generator?

3. Explain why the bacteria first generated carbon dioxide gas before producing the methane gas.

4. Where did the microorganisms that produced the methane gas come from in this experiment?

Name ______________________________ Date ______________

5. Describe why biogas is considered to be a biotechnology.

6. What are two advantages that biogas generation might have for our society?

7. What are other fuels that might be produced by living organisms?

Name ______________________________ Date ______________

CHAPTER 14
Consumer Concerns about Biotechnology

LAB ACTIVITY 14.1
Bioremediation Part 2: Tertiary Wastewater Treatment

Purpose

In this activity, you will investigate the applications of biotechnology in wastewater treatment. This time, instead of microorganisms, the use of macroorganisms, like plants, will be explored. As shown in Lab Activity 13.1, secondary wastewater treatment is very effective in removing unwanted, dissolved biodegradable substances from wastewater. Nutrient pollutants, such as nitrates and phosphates, however, remain in the water. Although water quality legislation does not require that these nutrient pollutants be removed from wastewater, research shows the potential harm (eutrophication and algal bloom) that they can do to aquatic environments. Because nitrates and phosphates are indeed nutrients to plants, it seems plausible that a wastewater treatment system can be designed to use plants as a means to filter out the unwanted nutrients.

Materials

- Simulated wastewater (chlorine-free water mixed with a weak solution of water-soluble fertilizer)
- Water hyacinths or other aquatic plants and algae
- Aquarium gravel
- Aquarium aerator pump
- Plastic tubing
- Clear containers (plastic soda bottles or small aquarium tanks)
- Utility knives
- Nitrate test kit
- Phosphate test kit
- Graph paper or computer spreadsheet application
- Colored pencils

Procedure

1. Collect all the required materials and bring them to your work area. Use the diagram in Lab Figure 14–1 to help you construct your tertiary wastewater treatment tank. If you are using a plastic soda bottle, use the utility knife to carefully cut off the top third of the bottle. The bottom part of the bottle will act as your wastewater treatment tank. If you are using aquatic plants other than water hyacinths, then you must line the bottom of the tank with gravel to help hold down the plants' roots. Because water hyacinths float on the surface of the water, they do not require gravel.

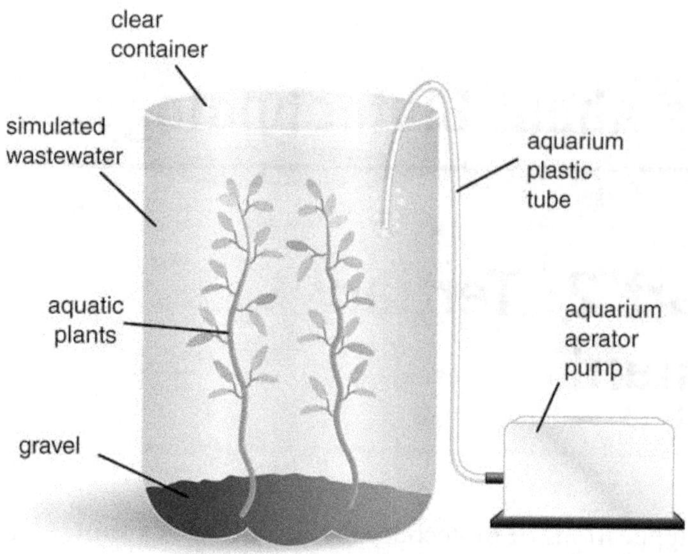

Lab Figure 14–1
Tertiary Wastewater Treatment Model

2. Once your tank is prepared, carefully fill it with the simulated wastewater. Then add your aquatic plants.
3. Next, you must aerate the water with the aquarium aerator pump and tubing.
4. Finally, using the test kits provided and with help from your instructor, test the wastewater for its nitrate and phosphate concentrations. Record the measurements in the format shown in Lab Figure 14–2.
5. Continue to make daily measurements of the nitrate and phosphate levels of the wastewater for a period of 2–3 weeks until your experiment has concluded.
6. Once your experiment is completed, construct a multiple line graph that shows the change in concentration of both nitrate and phosphates as recorded in the format shown in Lab Figure 14–2. The *x*-axis should be labeled "day." The primary *y*-axis should be labeled "nitrate concentration" and the secondary *y*-axis "phosphate concentration." Use a different-colored pencil for each line and create a key for your graph.

Name____ Date__________________

Date	Nitrate Concentration	Phosphate Concentration

Lab Figure 14–2

Name ______________________________ Date ______________

Conclusions

1. Briefly summarize the results of your experiment.

____

____

____

____

____

2. Describe how this experiment illustrates the application of biotechnology to wastewater treatment.

____

____

____

____

____

3. Explain how you could incorporate tertiary wastewater into existing wastewater treatment plants.

____

____

____

____

____

4. Describe some of the problems that might arise by adding tertiary treatment to all wastewater treatment plants.

____

____

____

____

____

5. Calculate the daily rate of change for both nitrate and phosphate concentration over the length of your experiment. Rate of change can be calculated by using the formula: rate of change = change in concentration/change in time.

____

____

____

____

____

6. Explain the importance of tertiary wastewater treatment.

____

____

____

____

____

Name______________________________ Date______________

LAB ACTIVITY 14.2
Antibacterial Properties

Purpose

The purpose of this lab is to test the antibacterial effects of various household materials.

Materials

- *Escherichia coli* bacteria culture
- Petri dishes
- Nutrient agar
- Sterile loops
- Laboratory film
- Antibiotic disks
- Forceps
- Various household items with possible antibacterial properties. Ideas include the following:
  - Bleach
  - Antibacterial soaps
  - Dish soaps
  - Cinnamon
  - Vinegar
- Distilled water (if using nonliquid items)
- Lab incubator (optional)

Procedure

1. Pour ½ inch of nutrient agar in the bottom of each Petri dish. Allow the agar to solidify.
2. Each lab group should have three Petri dishes with agar and three sterile loops.
3. Use a sterile loop to transfer a colony of bacteria from the bacteria culture to a Petri dish. Use the loop to spread the bacteria over the surface of the agar, starting at the edge of the plate and moving toward the center. Turn the dish 45°, and repeat. Continue this process until the entire surface of the agar has been covered.
4. Repeat step 3 for the other two Petri dishes. Be sure to use a different sterile loop for each dish.
5. Obtain eight antibiotic disks. Use the forceps to dip four of the disks in one type of house hold material. Place the disks in a circle in the middle of one of the dishes.
6. Repeat step 6 with a different household material.
7. Do not put any disks in the third dish. The Petri disk without the disks will serve as your control.
8. Cover the Petri dishes and wrap them with laboratory film to prevent contamination.
9. Place the Petri dishes in a laboratory incubator set at 37°C for 36–48 hours.
10. After incubation, compare the bacterial growth in each dish.

Name______________________________ Date__________

Conclusions

1. Draw a diagram of each dish. Be sure to label the location of the disks.

2. Based on your results, what can you conclude about the antibacterial properties of the house hold products you tested?

3. What other household products would you like to test for antibacterial properties? Why would you test these items?

4. What are the benefits of using products with antibacterial properties in your home? Can you think of any drawbacks of using these products?

Name______________________________ Date______________

CHAPTER 15
Ethical Issues in Biotechnology

LAB ACTIVITY 15.1
Bioremediation Part 3: Oil-Degrading Bacteria

Purpose

In this activity, you will explore the potential for using microorganisms to clean up oil spills. You will conduct two experiments to test the possibility of using organisms as a means to clean up damage done to the environment by petroleum compounds.

Materials

- Oil-degrading microbes
- Two 500-ml beakers
- Aquarium pump
- Plastic aquarium tubing
- T-joint air splitter for aquarium tubing
- 6 grams oil
- 400 ml distilled water
- Two sterile Petri dishes containing nutrient agar, with lids
- Sterile swabs
- Sand
- Safety goggles
- Rubber gloves
- Two disposable eyedroppers
- Biohazard autoclave bag
- Plastic wrap
- Tape

Caution: Follow all of the directions as outlined by your instructor about the safe handling of the bacterial and fungal cultures used in this experiment.

Procedure A

1. Label one beaker "A" and the other beaker "B." Fill each beaker with 200 ml of distilled water. Add 3 grams of oil to the water in each beaker. This will represent an oil spill in an aquatic environment.
2. Label beaker A "no oil-degrading microbes" and beaker B "oil-degrading microbes."
3. Wearing your safety goggles and gloves, use a disposable eyedropper to obtain a sample of oil-degrading microbes from your instructor. Carefully add the microbes into the water and oil mixture in beaker B. Dispose of the dropper in the biohazard bag.

4. Set up your aquarium pump and tubing and place a tube into each beaker so that the tubes are below the surface and the pump is aerating both mixtures of oil and water.
5. Cover each beaker loosely with plastic wrap to prevent excess evaporation of the water.
6. Observe the appearance of both oil and water mixtures over the next 3–4 weeks.

Procedure B

1. Obtain two sterile Petri dishes from your instructor. Do not open them. Keep the lids on the dishes at all times. Mark the bottom of one of the Petri dishes with an "A" and the other one with a "B." Also mark each dish with your name.
2. Put on your gloves and safety goggles. Using a sterile applicator swab, obtain a sample of the oil-degrading microbes culture from your instructor. Quickly open Petri dish B. Inoculate the nutrient agar by gently rubbing the swab containing the microbes on the growth medium, using the technique shown by your instructor. Quickly replace the lid of the dish.
3. Dispose of the swab with the oil-degrading microbes in the biohazard autoclave bag provided by your instructor.
4. Using a disposable eyedropper, obtain a sample of oil and put four to five drops of oil into each Petri dish. Quickly replace the lid of each dish.
5. Seal the sides of each Petri dish with tape and place both dishes in a dark place designated by your instructor.
6. Observe both dishes over the period of a few weeks. Sketch your observations for each dish within circles labeled "A: no microbes" and "B: oil-degrading microbes" (see Lab Figure 15–1).

Name ______________________________ Date ______________

Date	Petri Dish A: No Microbes	Petri Dish B: Oil-Degrading Microbes

Lab Figure 15–1

Name____ Date______________

Conclusions

1. Explain why you had to use two beakers of oil and water for Procedure A.

2. Describe the results of the experiment you conducted in Procedure A.

3. Did you observe any differences in the appearance of Petri dishes A and B in Procedure B? If so, describe these differences.

4. Which experiment yielded the quickest results, and why?

Name____________________ Date__________

5. Did this activity show the benefits of using organisms to clean up spilled oil?

____

____

____

____

____

6. Explain how you would use oil-degrading microorganisms to clean up a real oil spill.

____

____

____

____

____

Name________________________________ Date____________

LAB ACTIVITY 15.2
Toxicology Lab

Purpose

The purpose of this lab is to determine the toxic effects of common substances on a plant species.

Materials

- Seedlings or seeds of various types (each student will need 10 seedlings or seeds)

> ***Tip:*** Fast-growing plants are the best for this kind of experiment.

- Potting soil
- Pots
- Graduated cylinder
- Beakers
- Measuring spoons
- Distilled water
- Household materials to test for toxicity. Ideas include the following:
  - Vinegar
  - Salt
  - Bleach
  - Soda
  - Fertilizer

> ***Tip:*** Do not use any material that will smell when it is left unrefrigerated.

Procedure

1. Determine what substance you will test for toxicity and what plant you will use as the test subject.
2. Determine what dose will be considered the 100 percent dose or the maximum dose.
3. If your substance is not in liquid form, you will need to mix your 100 percent dose with 100 ml of distilled water.
4. Create a serial dilution of your toxic substance by taking 10 percent of your 100 percent dose and mixing it with 90 percent distilled water to create a 10 percent dose. Then repeat the process to create a 1 percent dose and a .1 percent dose.
   a. For example, if your 100 percent dose is 100 ml of substance, do the following:
      i. To create a 10 percent solution, mix 10 ml of your original substance with 90 ml of distilled water.
      ii. To create a 1 percent solution, mix 10 ml of the 10 percent solution with 90 ml of distilled water.
      iii. To create a .1 percent solution, mix 10 ml of the 1 percent solution with 90 ml of distilled water.

5. Plant 10 seeds or seedlings in five pots—two plants or seeds per pot. Be sure the soil in each pot is well moistened.
6. Label the pots "100%," "10%," "1%," ".1%," and "control."
7. Pour 100 ml of the 100 percent solution into the pot labeled "100%," 100 ml of the 10 percent solution into the pot labeled "10%," and so on. Add 100 ml of distilled water to the pot labeled "control."
8. Set your pots in an area that receives adequate warmth and sunlight for plant growth. Be sure all your pots are in the same location.
9. Set up a watering schedule for your plants. Always water your experimental plants with your test solution and always water your control plant with distilled water.
10. Record observations of your plants daily for 3 weeks.

Name________________________________ Date______________

Conclusions

1. What household material did you test for toxicity? What can you conclude about the toxicity of your substance based on your observations?

2. Why do you need a control for this experiment?

3. What is the purpose of using a serial dilution to create the different doses of the test substance?

4. Toxicologists call the dose at which 50 percent of organisms die from exposure to a substance the lethal dose 50 percent or LD50. Can you determine the LD50 of your substance?

5. How would you change your experiment to get a more accurate estimate of the LD50?

Name____ Date______________

CHAPTER 16
Careers in Biotechnology

LAB ACTIVITY 16.1
Phytoremediation

Purpose

In this activity, you will investigate the benefits of using plants as a way to clean up environmental pollutants such as toxic soils. This process is called phytoremediation and is an exciting new aspect of biotechnology. You will screen various plants to identify each one's potential as a hyperaccumulator, which is a plant that is capable of taking up large concentrations of toxic metals from polluted soil.

Materials

- Potential hyperaccumulating plants, such as alpine pennycress (*Thlaspi caerulescens*), field mustard (*Brassica rapa*), red fescue (*Festuca rubra*), Indian mustard (*Brassica juncea*), geraniums (*Pelargonium*), and common sunflower (*Helianthus annuus*)
- Two 2-liter plastic soda bottles with caps
- Utility knife
- Two 5-inch-long candlewicks or thick cotton strings
- Growth medium (1:1 ratio of perlite and peat moss)
- Distilled water
- 1,000-ml beaker
- .1 M zinc nitrate
- Voltage meter
- Table salt
- Eyedropper
- Stirring stick
- 1-inch section of a plastic straw
- Tape
- Graph paper or spreadsheet computer application
- Colored pencils

Caution: Always wear your safety goggles and rubber gloves and observe all safety procedures as outlined by your instructor when handling the zinc nitrate solution.

Procedure A: Building Your Growth Chambers

1. To begin this activity, you will construct two plant growth test chambers. This will be done by modifying two 2-liter plastic soda bottles. The growth chambers used in this experiment have been adapted from the University of Wisconsin's Bottle Biology program.

2. Using the utility knife, carefully cut off the tops of the two soda bottles, approximately 5 inches from the bottom of each bottle.
3. Next, with help from your instructor, bore a small hole into the top of the cap of each bottle.
4. Now, stick the cotton wicks through the hole in each cap and screw the caps onto the bottle tops. Approximately half the wick should be through the hole in the cap.
5. Invert the tops of the soda bottles into the bottoms of the bottles, which act as reservoirs for your test solution.
6. Add approximately 400 ml of growth medium into the top portion of each bottle, making sure that the cotton wick is suspended in the middle of the soil.
7. Next, cover the reservoir portion of each plant growth chamber with a piece of paper and tape it into place. This will help to reduce the growth of algae in your test solution. Your plant growth chambers are now complete and ready for use. They should resemble the one shown in Lab Figure 16–1.
8. With help from your instructor, obtain two different species of potential hyperaccumulating plants. Carefully shake off any excess soil from the roots of the plants and plant one plant in each growth chamber. Moisten the soil with a small amount of distilled water. Label each growth chamber with your name and the name of the species of plant being tested.

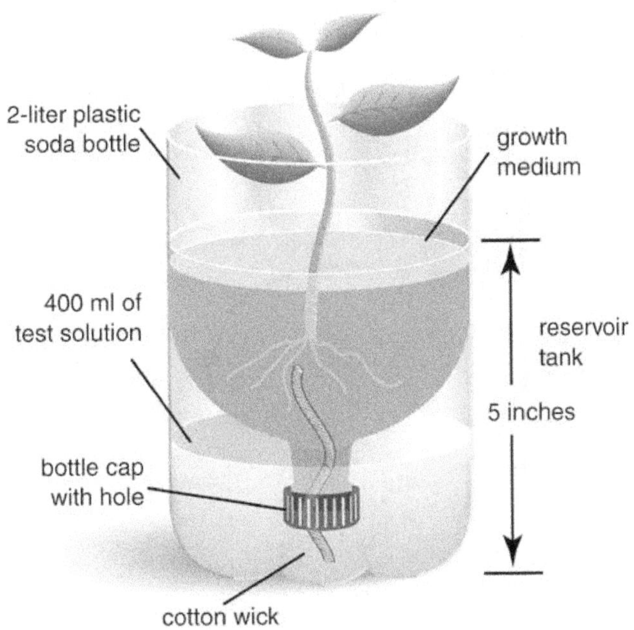

Lab Figure 16–1
Plant Growth Test Chambers for Hyperaccumulating Plants

Name____ Date_______________

Procedure B: Constructing a Test Meter and Mixing Your Test Solution

1. Now you will make a solution that will be used to test the hyperaccumulating abilities of your plants. First, you will construct a simple solution concentration meter, using a voltage meter. The voltage meter will measure the relative concentration of the metals in your test solution by testing the electrical conductivity of the solution. Conductivity is a measure of how much electricity can be conducted by a substance, in this case a solution of zinc nitrate. The greater the concentration of metals in a solution, the greater the electrical conductivity. The electrical conductivity of a solution can be measured by determining the electrical resistance of the solution. The greater the resistance of a solution, the lower the concentration of metals in the solution. The lesser the resistance, the higher the concentration of metals in the solution. A voltmeter measures resistance by using ohms. Therefore, a solution that has a high ohms reading, or high resistance, has a low concentration of metals. A low ohms reading, or low resistance, has a high concentration of metals. By monitoring the changes in the resistance of your test solution, it is possible to make relative measurements of the uptake of metals by your plants.
2. To make a simple concentration meter, take a 1-inch section of a plastic straw and tape each tip of the voltmeter's probes to each end. This will act as a simple test probe that will calibrate your meter. The straw keeps each probe at a constant distance from each other, thereby calibrating the reading, see Lab Figure 16–2.

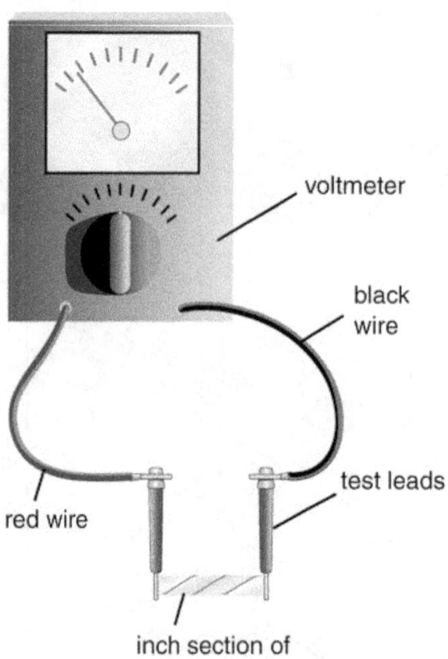

Lab Figure 16–2
Simple Test Meter for Measuring the Concentration of Test Solution

3. Next, pour about 200 ml of distilled water into a beaker. Place your test probe into the water. Distilled water has no electroconductivity, so the voltmeter should measure a very high resistance or high ohms reading.
4. Now carefully add a small amount of table salt into the beaker containing the distilled water. As you add the salt, the voltmeter should begin to measure the conductivity of the solution. Because the resistance is decreasing, the ohms reading should begin to decrease. This is how you will measure the concentration of zinc in your test solution.
5. Now that your test meter is ready, discard the distilled water and salt solution from your beaker. Rinse out the beaker and now fill it with 800 ml of distilled water. Put on your safety goggles and rubber gloves and obtain the zinc nitrate stock solution from your instructor. You will now create a test solution that contains zinc, which you will use to test the hyperaccumulating abilities of your two plants. Place your test probe into the distilled water and, using an eyedropper, slowly add the zinc nitrate one drop at a time into the distilled water. After each drop, stir the water to mix the zinc nitrate into the water. Monitor the resistance measure by your voltmeter after you add each drop. You want to add enough zinc nitrate so that your meter is measuring a mid- to low-range ohm reading. This will create a test solution with a mid- to high-range concentration of zinc. Once you have reached the desired concentration of zinc nitrate in your test solution, record the electrical resistance in ohms of your solution in the format shown in Lab Figure 16–3. Your solution is now ready for use.

Procedure C: Zinc Hyperaccumulating Plant Screening

1. Pour 400 ml of the test solution you created in Procedure B into the reservoirs of each plant growth chamber you built in Procedure A. You will now monitor the growth of each plant and the concentration of the zinc in the test solution every day for the next few weeks. If your plants begin to uptake and hyperaccumulate the zinc from your solution, then your voltmeter will begin to measure an increase in your ohms reading as the electrical resistance of the solution increases and the conductivity decreases as a result of the loss of zinc. Record each daily ohms reading in the format shown in Lab Figure 16–3 for each plant you are testing.
2. At the conclusion of your experiment, use the data you recorded to create two line graphs that show the change in electrical resistance over the period of your experiment. This change will illustrate the uptake of zinc by each plant and should help you identify if the plant is an effective hyperaccumulator. Label your x-axis "date" and y-axis "electrical resistance in ohms."

Name ______________________________ Date ______________

Date	Plant A: Electrical Resistance (ohms)	Plant B: Electrical Resistance (ohms)

Lab Figure 16–3
Hyperaccumulating Plants Test Screening

Name ____________________ Date ____________

Conclusions

1. Define the terms bioremediation, phytoremediation, and hyperaccumulating plants and explain how they relate to biotechnology.

2. What are the advantages of using plants to clean up environmental pollution?

3. What are the disadvantages of using plants to clean up environmental pollution?

4. What two potential hyperaccumulating plants did you use in your experiment?

5. Explain how electroconductivity can be used to measure the concentration of metals in a solution.

Name____ Date ______________

6. Describe the results of your experiment.

____

____

____

____

7. Calculate the rate of change in electrical resistance for both of your plant's test solutions over the course of your experiment using the formula: rate of change = change in resistance/number of days.

____

____

____

____

8. Which of your plants did you find to be the most effective hyperaccumulator, and why?

____

____

____

____

9. Did the results of this experiment support the possibility of using plants to clean up toxic metals in the soil? Explain.

____

____

____

____

Name____ Date______________

LAB ACTIVITY 16.2
Design Your Own Inquiry Lab

Purpose

The purpose of this lab is to design your own scientific study based on the scientific method.

Materials

- Access to the Internet, library or other research materials
- Poster board or trifold boards
- Laboratory notebook
- Various other lab supplies as required by projects

Procedure

1. Pick a topic in biotechnology that interests you. Be sure it is a topic that you can realistically explore given the conditions and setting of your classroom and the materials available to you.
2. Research your topic and determine a question you want to answer through scientific testing.
3. Write a hypothesis based on your question.
4. Design an experiment to test your hypothesis. Be sure to base your experimental design on the scientific method. Get your teacher's approval before proceeding to the next step.
5. Gather your materials and set up your experiment.
6. Collect data and make daily observations. Be sure to record all pertinent information in your laboratory notebook. Continue to record data for the length of your experiment.
7. On conclusion of your experiment, analyze your data and draw conclusions.
8. Create a poster to display your research, question, hypothesis, experimental design, data, and conclusions. Share your poster with the other students in your class.

Tip: The laboratory notebook and poster resulting from this lab can be used for the FFA Agriscience CDE. Be sure to read and follow the CDE rules posted at http://www.ffa.org.

Name ______________________________ Date ______________

Conclusions

1. What was the question you were trying to answer?

2. In your experimental design, what were your controls and your variable(s)?

3. Why is it important to have controls when completing an experiment?

4. Did your data support or refute your hypothesis? Give evidence.

Name ______________________________ Date ______________

5. What are some sources of error in your experiment?

____

____

____

____

____

6. If you could repeat this experiment, would you do anything differently? Why or why not?

____

____

____

____

____

NOTES

NOTES

NOTES

NOTES

NOTES

CPSIA information can be obtained
at www.ICGtesting.com
Printed in the USA
FFHW01n1349130818
47743737-51423FF